THE ULTIMATE LOAD DESIGN
OF
CONTINUOUS CONCRETE BEAMS

THE ULTIMATE LOAD DESIGN OF CONTINUOUS CONCRETE BEAMS

DERRICK BECKETT
B. Sc. (Eng.) (Lond.), D. I. C., A. M. Inst. H. E.

Lecturer in Civil Engineering, University of Surrey

Springer Science+Business Media, LLC

Suggested U.D.C. numbers: 624.012.4
624.072.2
624.042

Library of Congress Catalog Card Number 67–31269

ISBN 978-1-4899-6160-0   ISBN 978-1-4899-6297-3 (eBook)
DOI 10.1007/978-1-4899-6297-3

PREFACE

This book is intended for the use of final year undergraduates
and in the design office. It is an attempt to bridge the gap between conventional design procedures for continuous concrete
beams, based on an elastic distribution of moments at working
load, and the ultimate load approach, which is based on the
plastic behaviour of concrete beams. The text assumes that the
reader has an elementary knowledge of the elastic sizing of
reinforced concrete sections and is reasonably familiar with
the section of CP 114 : 1957 relating to design considerations.
Chapter 1 discusses the limitations of elastic design and the
basis of ultimate load design together with British, American
and Russian design rules. Chapters 2 and 3 deal with the behaviour of reinforced and prestressed concrete sections under
load. In Chapter 4 the formation of plastic hinges is discussed
together with a summary of the work of Professor A.L.L.
Baker. The final two chapters consider serviceability requirements, crack width limitations etc., and the problems of shear
and torsion. Several worked examples are included in the text
together with a comprehensive reference list and metric con-
version table.

The author would like to express his thanks to P. R. Knowles,
M. A. (Cantab.), A.M.I.C.E. for his encouragement and help
in the preparation of the text.

CONTENTS

PRINCIPAL NOTATION

A_{sc} Area of compressive reinforcement
A_{st} Area of tensile reinforcement
b Width of section
C_c Crushing strength of concrete cylinders
C_u Crushing strength of concrete cubes
d Effective depth of section
E_c Elastic modulus of concrete
E_s Elastic modulus of steel
e Eccentricity of prestress force
e_c Strain in concrete
e_s Strain in steel
f_{cb} Permissible compressive stress for concrete in bending (p_{cb}, CP 114)
f_{st} Permissible tensile stress in steel (p_{st}, CP 114)
f_{yc} Yield stress of steel in compression
f_{yt} Yield stress of steel in tension
l_a Lever arm
M_u Ultimate moment of resistance
nd Depth from top fibre to neutral surface
P Prestress force
p Steel percentage $= A_{st}/bd$
Q Resistance moment factor
w_d Dead load per unit length
w_s Superimposed load per unit length
w_t Total load per unit length $= w_d + w_s$
Δ Crack width

Symbols used in formulae quoted from building codes etc. are defined separately in the text.

METRIC CONVERSIONS

0·1	mm	= 0·004 in.
0·2	mm	= 0·008 in.
0·3	mm	= 0·012 in.
1·0	mm	= 0·04 in.
1·0	cm	= 0·4 in.
2·54	cm	= 1·0 in.
0·305	m	= 1 ft.
0·914	m	= 3 ft.
30·5	m	= 100 ft.
305	m	= 1,000 ft.
1·61	km	= 1 mile = 5,280 ft.

1 lb. (avoirdupois) = 0·454 kg
1 ton (2,240 lb.) = 1·016 tonne
1 in⁴.   = 41·62 cm⁴

1 lb./in.² =	0·000 703	kg/mm²
=	0·0703	kg/cm²
100 lb./in.² =	7·03	kg/cm²
200 lb./in.² =	14·1	kg/cm²
500 lb./in.² =	32·5	kg/cm²
1,000 lb./in.² =	70·3	kg/cm²
2,000 lb./in.² =	141	kg/cm²
3,000 lb./in.² =	211	kg/cm²
4,000 lb./in.² =	281	kg/cm²
5,000 lb./in.² =	352	kg/cm²
10,000 lb./in.² =	703	kg/cm²
20,000 lb./in.² =	1,406	kg/cm²
40,000 lb./in.² =	2,812	kg/cm²
60,000 lb./in.² =	4,219	kg/cm²
29,000,000 lb./in.² =	2,039,000	kg/cm²

1

LIMITATIONS OF ELASTIC DESIGN

This book is concerned with recent developments in the design of continuous concrete beams. In the past fifteen years there has been a marked change in the approach to the design of concrete structures, both in terms of structural conception and the sizing of individual members. With regard to structural conception the use of skeletal frameworks for the multistorey residential and office developments has lost favour due to the planning limitations of large column and beam sizes and the need for providing the maximum number of floors in the minimum possible height of construction. More efficient use has been made of the material in the structure such as gable, lift and stair walls to carry wind load, the floors acting as deep horizontal beams. This has led to reduction in column sizes as they can be designed for predominantly direct loading. The beams, not being subjected to wind load moments, can also be reduced in size. A further important consideration is the increasing complexity of service requirements as the need for an improved working environment grows. In order to obtain maximum room for services (pipes, warm air ducts etc.) whilst maintaining the maximum number of floors for a given height of structure, the structural engineer is being pressed to minimize construction depths. At the same time an adequate factor of safety must be maintained and a high standard of serviceability at working load is required in terms of deformation and crack widths.

Few engineers would now design a series of continuous beams on the assumption of an elastic distribution of moments. This would lead to uneconomic structural depths and it is interesting to note that the elastic design of reinforced concrete assuming a constant ratio m of the modulus of elasticity of steel to that of concrete, was abandoned in Russia in 1938. The British Standard Code of Practice, CP 114, introduced the so called 'load factor' method of sizing reinforced concrete sections for bending in 1957, and also allows a 15 per cent adjustment to moments obtained by elastic analysis, which implies inelastic behaviour. The Institution of Civil Engineers appointed a committee in 1958 to review existing knowledge with regard to the ultimate load design of concrete. The findings of this committee, which worked in close liaison with the European Concrete Committee (C.E.B.) were published in 1962[1]. Draft recommendations[2] for an international code of practice have now been published which give basic equations for ultimate load design. Further recommendations for the ultimate load design of reinforced concrete structures[3] were presented for discussion at a joint meeting of the Cement and Concrete Association, the Institution of Civil Engineers, the Institution of Structural Engineers and the Reinforced Concrete Association in March 1965. Before considering the ultimate load design of reinforced concrete beams the conventional approach in accordance with CP 114 (1957) will be briefly reviewed. An elastic distribution of bending moments may be obtained by moment distribution methods etc., the stiffness factors being based generally on uncracked sections. A limited transfer of bending moments is permitted between critical sections of up to 15 per cent. The designer is then given the alternative of sizing the concrete section assuming a modular ratio of 15 or applying the 'load factor' equations which reduce the stress/strain conditions at ultimate load conditions to an equivalent relationship at working load. In order to familiarize the reader with the two methods, the relevant equations will be developed from first principles for a singly reinforced section. This will also enable the two methods to be compared quantitatively.

ELASTIC SIZING OF A SINGLY REINFORCED SECTION

Consider a rectangular section of width b, effective depth d (*Figure 1a*) with reinforcement area A_{st}. The modular ratio is m and the permissible concrete and steel stresses are f_{cb} and

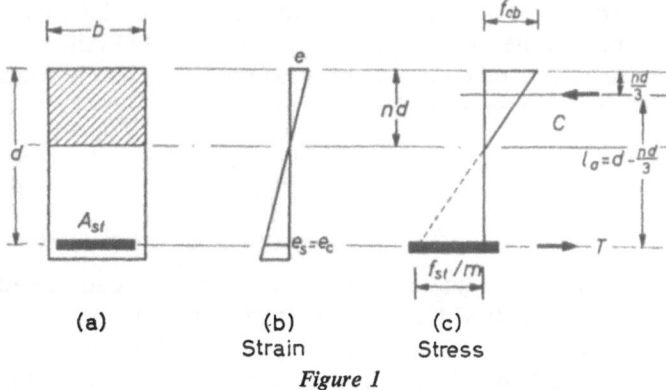

(a) (b) Strain (c) Stress

Figure 1

f_{st} respectively. The stress diagram is indicated in *Figure 1c*, the neutral surface being at a depth nd from the top fibres and the resistance of concrete to tensile stress is ignored. The modular ratio is obtained as follows:

Modulus of elasticity of steel

$$E_s = \frac{\text{Steel Stress}}{\text{Steel Strain}} = \frac{f_{st}}{e_s}$$

Modulus of elasticity of concrete

$$E_c = \frac{\text{Concrete Stress}}{\text{Concrete Strain}} = \frac{f_{cb}}{e_c}$$

At the level of the steel $e_s = e_c$ (*Figure 1b*).

Hence
$$\frac{f_{st}}{E_s} = \frac{f_{cb}}{E_c}$$

$$f_{st} = \frac{E_s}{E_c} f_{cb}$$

$$= m f_{cb} \quad \text{where} \quad m = \frac{E_s}{E_c}$$

Referring to *Figure 1c* the equivalent stress in the concrete at the level of the steel is f_{st}/m. According to CP 114 the modular ratio can be taken as 15 and thus the depth to the neutral surface is obtained from similar triangles

$$\frac{f_{cb}}{\dfrac{f_{st}}{m}} = \frac{nd}{d-nd} = \frac{n}{1-n} \qquad (1.1)$$

Thus for given values of f_{cb} and f_{st} the value of n can be determined from equation 1.1. Considering the longitudinal equilibrium of the section, the total compression C is equal to the total tension T, and these two forces provide the internal moment of resistance of the section to balance the applied moment.

The resistance moment $= Cl_a = Tl_a$

where
$$l_a = d - \frac{nd}{3} = d \left(1 - \frac{n}{3} \right)$$

$$Cl_a = \frac{f_{cb}}{2} bnd = d \left(1 - \frac{n}{3} \right)$$

$$= Q \, bd^2 \qquad (1.2)$$

where
$$Q = \frac{f_{cb}}{2} n \left(1 - \frac{n}{3} \right)$$

Hence for elastic design the resistance moment can be expressed in the form $RM = Qbd^2$ where Q is a factor depending on the allowable stress in the concrete and the depth to the neutral surface. The value of n is related to the allowable steel and concrete stresses and the modular ratio m. This leads to the first limitation of the elastic concept of design, that is the value chosen for m. As the modular ratio can vary from about 5 to 30, according to the crushing strength, choice of the value of 15 is rather arbitrary. Further, due to the creep of concrete under load, that is change in strain at a constant stress, the effective modulus of elasticity of the concrete will differ from the elastic modulus. A section sized using equations 1.1 and 1.2 will generally give adequate serviceability but no indication is given of its factor of safety against failure.

'LOAD FACTOR' SIZING OF A SINGLY REINFORCED SECTION

Use of equations 1.1 and 1.2 implies that the stresses substituted in these equations relate to the elastic state of stress for both steel and concrete. Consider now typical stress-strain characteristics for mild steel and concrete as indicated in *Figure 2*. The stress-strain curve for mild steel has a definite straight line portion OL_1 which represents the elastic range. At a stress f_{yt} the steel yields and the portion L_1L_2 represents the inelastic range in which the steel undergoes extensive strain with very little increase in stress. The dotted line represents a simplified form of this curve consisting of two straight lines OL_1 and L_1L_2, which is sometimes referred to as the idealized stress-strain diagram. The concrete curve has no straight line portion but again can be represented in idealized form by the dotted line OL_1L_2.

Similarily an idealized stress-strain curve can be drawn for high tensile steel. The limiting concrete strain L_2 is generally in the range $0·003 - 0·004$ and for the present argument a mean value of $0·003\ 5$ *(Figure 3b)* will be assumed. The stress distribution at the ultimate condition will be as indicated in *Figure 3c*. Yield of mild steel takes place at a strain value of about 0·002 and if the percentage of steel in the section is such that the steel will yield before the concrete fails in compression the section is referred to as under-reinforced. With this type

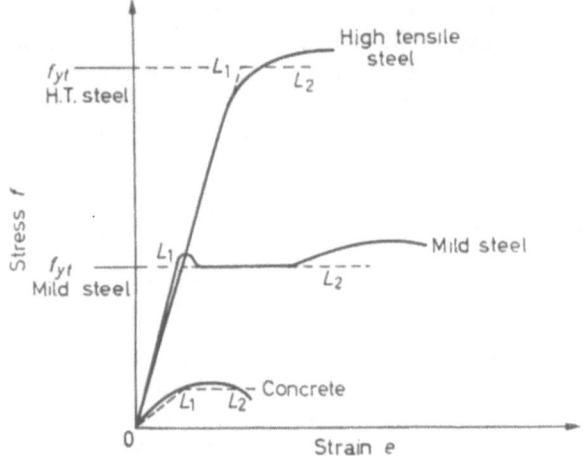

Figure 2. Actual and idealized stress-strain curves for steel and concrete

of section there is ample warning of failure, because at a load somewhat less than the ultimate, the steel will yield, and further increase in load will cause extensive cracking, the neutral axis rising until crushing of the concrete occurs. This gradual form of failure is obviously preferable to sudden failure which would occur in section in which the percentage of steel was such that the concrete crushed before the steel reached its yield stress. This is referred to as an over-reinforced section. Considering *Figure 3a* and assuming the area of steel is such that the yield stress f_{yt} is developed at the limiting concrete strain of 0·003 5, then this represents the so-called balanced state. At this condition the depth of the neutral surface is given by

$$\frac{n}{l-n} = \frac{0\cdot003\ 5}{0\cdot002}$$

$$n = \frac{0\cdot003\ 5}{0\cdot005\ 5} = 0\cdot637$$

To ensure an under-reinforced section, n would have to be less than this value and in fact CP 114 gives a limiting value of 0·5. The maximum concrete stress at failure *(Figure 3c)* is generally taken as approximating to a constant compressive stress of two thirds of the cube strength, C_u. In CP 114 it is further stipulated that due to the greater variability of the strength of concrete compared with steel, the cube strength of concrete should be taken as two-thirds the actual cube strength so that the section is subjected to a constant stress of 4/9 C_u over a depth not greater than $d/2$, at ultimate conditions. An expression can now be developed for the resistance moment of

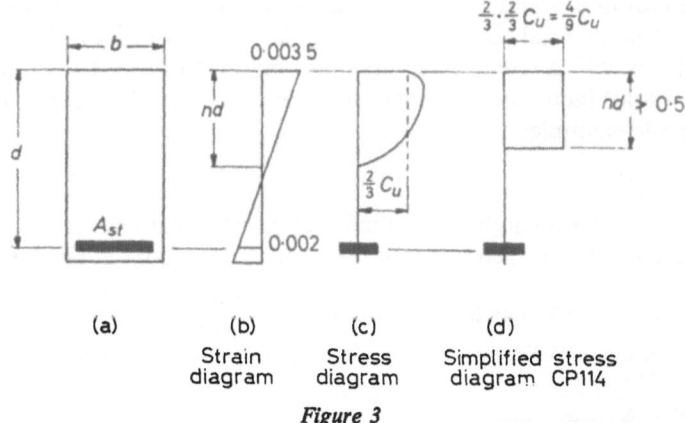

(a)	(b)	(c)	(d)
	Strain diagram	Stress diagram	Simplified stress diagram CP114

Figure 3

the section of similar form of equation 1.2. From *Figure 3d* for

$$n = 0\cdot5$$

$$RM = CL_a = Tl_a$$

$$= \frac{4}{9} C_u \frac{bd}{2} \cdot \frac{3}{4} d$$

$$= \frac{C_u}{6} bd^2 \tag{1.3}$$

Equation 1.3 represents the ultimate moment of resistance of the section and has to be divided by a factor of safety to give the working load moment of resistance. Assuming a factor of safety of 2 then the resistance moment is equal to $\frac{1}{12} C_u bd^2$. For $C_u = 3,000$ lb/in.^{2} $RM = 250 bd^2$, that is $Q = 250$. In CP 114 the resistance moment for working load is given by

$$RM = \frac{f_{cb}}{4} bd^2 \tag{1.4}$$

For $C_u = 3,000$ lb./in.^{2}, and $f_{cb} = 1,000$, then RM $250 bd^2$. Thus equations 1.3 and 1.4 give the same result, the former relating to ultimate load conditions and the latter to working load conditions with a factor of safety (load factor)* of $2\cdot0$. The elastic and load factor methods will now be compared by means of a simple example.

Example 1.1

Consider a singly reinforced section of width $b = 12$ in. subjected to an applied bending moment (obtained from an elastic

---

* Load factor may be defined as the ratio $\dfrac{\text{Ultimate load}}{\text{Working load}}$. In elastic design the term Factor of Safety is frequently used, which is the ratio $\dfrac{\text{Yield stress}}{\text{Working stress}}$.

analysis) of 600,000 lb. in. The section will be sized elastically and by the ultimate load method for $C_u = 3,000$ lb./in.², that is $f_{cb} = C_u/3 = 1,000$ lb./in.² according to CP 114. Using mild steel reinforcement $f_{yt} = 40,000$ lb./in.² and $f_{st} = 20,000$ lb./in.²

(a) Elastic sizing

From equation 1.1

$$\frac{1,000}{\dfrac{20,000}{15}} = \frac{n}{1-n}$$

hence $$n = 0.428$$

From equation 1.2

$$Q = \frac{1,000}{2} \times 0.428 \times \left(1 - \frac{0.428}{3}\right) = 184$$

hence $$RM = 184bd^2$$

for $b = 12$ in. $$d = \left(\frac{600,000}{12 \times 184}\right)^{\frac{1}{2}}$$

$$= \underline{16.5 \text{ in.}}$$

$$l_a = d\left(1 - \frac{n}{3}\right) = 0.857$$

Thus $$A_{st} = \frac{600,000}{14.12 \times 20,000}$$

$$= \underline{2.125 \text{ in.}^2}$$

The above result gives no indication of the margin of safety against failure. This can be obtained by the consideration of equilibrium at ultimate load (*Figure 4a* and *b*)

$$C = T$$

$$\frac{4}{9}3,000 \times 12 \times nd = A_{st}f_{yt} = 2.125 \times 40,000$$

$$nd = 5.32 \text{ in.}$$

Ultimate moment

$$M_u = \frac{4}{9} \times 3{,}000 \times 12 \times 5.32 \left(16.5 - \frac{5.32}{2}\right)$$

$$= 1{,}178{,}000 \text{ lb. in.}$$

Hence the load factor for bending failure

is
$$\frac{1{,}178{,}000}{600{,}000} = 1.96$$

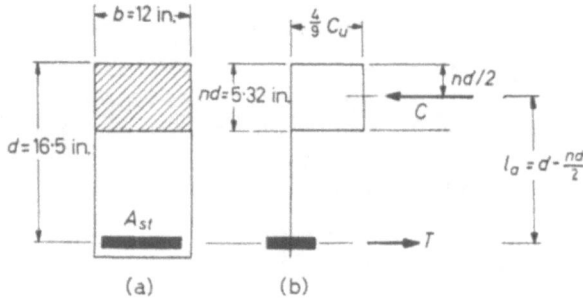

Figure 4

The elastic sizing for this particular case gives a factor of safety approximately to 2·0.

(b) Ultimate load sizing

From equation 1.3 $M_u = \dfrac{3{,}000}{6} bd^2 = 500bd^2$

The applied moment at working load = 600,000 lb. in.
For a load factor of 2·0, $M_u = 2 \times 600{,}000$ lb. in.

Thus
$$d = \left(\frac{1{,}200{,}000}{12 \times 500}\right)^{\frac{1}{2}}$$

$$= 14.12 \text{ in.}$$

The compression zone has been fully utilized,

that is $\qquad n = 0\cdot5 \qquad$ hence $\qquad l_a = 0\cdot75d$

and $\qquad A_{st} = \dfrac{1,200,000}{40,000 \times 0\cdot75 \times 14\cdot12}$

$\qquad\qquad = \underline{2\cdot83}$ in.^{2}

The ultimate load sizing gives a reduced construction depth but the steel area is increased. If the construction depth was limited to that obtained by the ultimate load method then an elastic sizing would require a doubly reinforced section as the resistance moment for $d = 14\cdot12$ in. would be less than 600,000 lb. in.

$$RM_{\text{elastic}} = 184 \times 12 \times 14\cdot12^2$$
$$= 441,000 \text{ lb. in.}$$

Where the depth of construction is not critical the ultimate load method of sizing will in some cases result in increased steel area[4] for singly reinforced sections. However, the above calculations were based on an elastic analysis of bending moments. Ultimate load sizing based on an elastic analysis of bending moments is somewhat illogical, and this leads to the consideration of the distribution of bending moments in continuous beams at ultimate load.

DISTRIBUTION OF BENDING MOMENTS IN CONTINUOUS BEAMS

Bending moment coefficients for an elastic analysis of a beam of constant cross section (subjected to uniformly distributed loading) continuous over four spans are indicated in *Figure 5*. The coefficients are given for all spans loaded *(Figure 5a)* representing dead load w_d, and for various live loadings w_s *(Figure 5b−e)* representing approximately maximum values at the supports and in the spans. CP 114 allows redistribution of

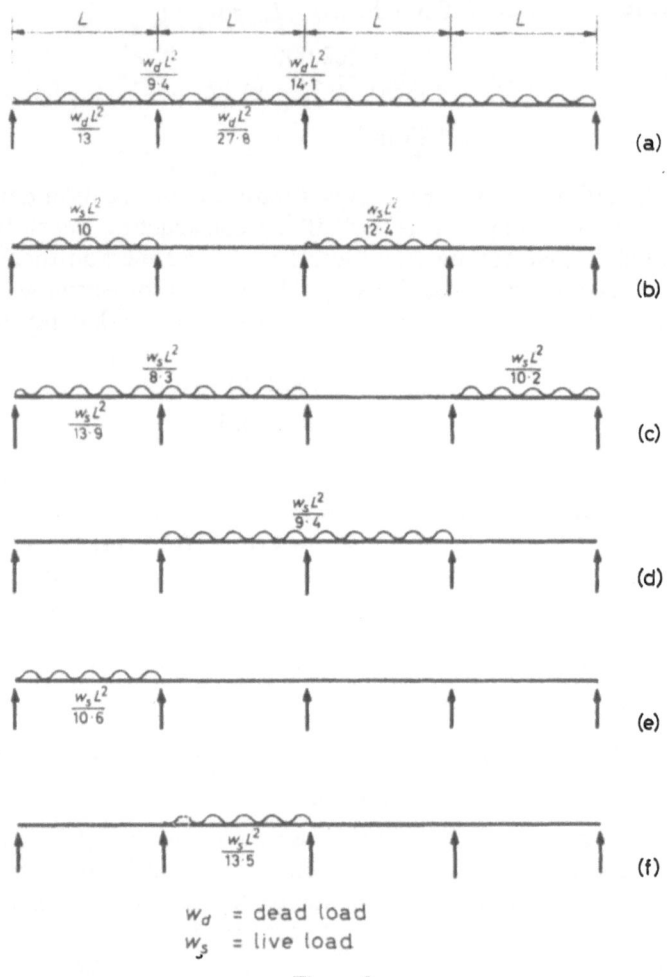

$$w_d = \text{dead load}$$
$$w_s = \text{live load}$$

Figure 5

bending moments indicated in *Figure 5* of up to 15 per cent, which implies inelastic behaviour. If a continuous beam is subjected to increasing load then at a certain stage of loading the steel will reach its yield stress at a particular section. If the beam is under-reinforced the section will have considerable plasticity, that is, it will be able to undergo considerable rota-

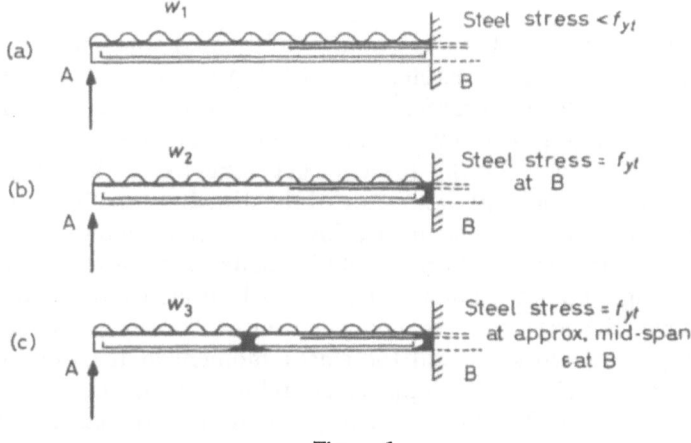

Figure 6

tion at approximately constant moment. Providing the limiting strain of the concrete is not exceeded further increase in load will be transmitted to adjacent sections. For example, in *Figure 6* the beam with loading w_1 is restrained at one end B and simply supported at A. Suppose the steel yields at B with loading w_2, then assuming adequate plasticity, point B can be considered as a plastic hinge or pivot. With further increase in load, AB may be taken as simply supported. As the load increases to w_3, a second hinge will form in the span, resulting in failure. This concept of plastic behaviour at ultimate load has been used extensively for the design of steel structures, but its use for the design of concrete structures has been restricted by doubt regarding the ability of reinforced or prestressed concrete sections to undergo the rotations implicit in a plastic hinge analysis.

There is however, considerable experimental evidence[1,3,5] to justify the plastic analysis of concrete beams, which produces a more economical distribution of bending moments and represents more closely the actual behaviour of the structure under load.

The behaviour of reinforced and prestressed concrete under load is amplified in Chapters 2 and 3 and the work of Professor A.L.L. Baker with regard to the calculation of permissible and actual hinge rotations is discussed in Chapter 4.

If we assume for the present that concrete sections have adequate plasticity at ultimate load the distribution of bending moments will differ considerably from those indicated in *Figure 5*. However a check should be made for serviceability at working load. For example a beam may be designed on an ultimate load basis for a moment coefficient at the support of say $w_t L^2/12$. At working load the elastic moment at the support may approach $w_t L^2/9$. Suppose the reinforcement has a yield stress of 60,000 lb./in.^{2} then at working load the stress will be 30,000 lb./in.^{2} for a load factor of 2·0. However, if we assume an elastic distribution of moment at working load the steel stress will be $30,000 \times 12/9 = 40,000$ lb./in.^{2} At this stress crack widths may be excessive. Hence an ultimate load design of a concrete structure requires a check on serviceability at working load in terms of crack widths and deformation. This is discussed further in Chapter 5.

So far bending only has been considered and the possibility of shear or torsional failure must be investigated. A member should be designed to ensure a gradual bending failure if full advantage is to be taken of plasticity, thus the proportioning of shear reinforcement should be such that failure due to overloading would occur in bending. Shear failure theories and methods of proportioning reinforcement are summarized in

Chapter 6. Before discussing in detail the behaviour of reinforced and prestressed concrete sections under load it will be
useful to compare moment coefficients recommended in various countries for the design of continuous beams of constant
cross section. Considering a continuous beam, then the British
code of practice states that for uniformly distributed loading
the following arrangements of live loading should be analysed.

(a) Alternate spans loaded and all other spans unloaded.

(b) Any two adjacent spans loaded and all other spans unloaded.

The moment coefficients obtained are indicated in *Figure 5*
and a 15 per cent adjustment is allowed. Alternatively, approximate bending moment coefficients for beams continuous over
three or more approximately equal spans are given and are
shown in *Figure 7*, Tables 1 and 2.

Two spans are taken as approximately equal if they do not
differ by more than 15 per cent of the longer span. Moment
redistribution is not allowed when using the approximate moment coefficients.

The American Concrete Institute Building Code (A.C.I.
318-63) gives moment coefficients as indicated in *Figure 7*,
Table 7. The longer of two adjacent spans should not exceed
the shorter by more than 20 per cent and the live loading should
not exceed three times the dead loading. For general building
work the live load is not likely to exceed three times the dead
load except for warehouse floors. Tables 3 to 6, *Figure 7*, give
moment coefficients for total load w_t ($w_t = w_s + w_d$) for the
ratio of live load to dead load equal to $\frac{1}{2}$, 1, 2 and 3 in accordance with the requirements of the British Code. These values
can be compared more easily with the A.C.I. coefficients. Russian recommendations[1] for moment coefficients are also given
in *Figure 7*, Table 8 in terms of the total load w_t. For other
continuous beams the design moments used should not be less
than 70 per cent of the moments obtained by elastic analysis
and the section should be sized so than n does not exceed 0·3,

Table	Loading	Near middle end span	Penultimate support	Middle interior spans	Other interior supports	Source
1	w_d	$\dfrac{w_d L^2}{12}$	$\dfrac{w_d L^2}{10}$	$\dfrac{w_d L^2}{24}$	$\dfrac{w_d L^2}{12}$	British code of practice
2	w_s	$\dfrac{w_s L^2}{10}$	$\dfrac{w_s L^2}{9}$	$\dfrac{w_s L^2}{12}$	$\dfrac{w_s L^2}{9}$	British code of practice
3	$\dfrac{w_s}{w_d}=\dfrac{1}{2}$	$\dfrac{w_t L^2}{11\cdot2}$	$\dfrac{w_t L^2}{9\cdot6}$	$\dfrac{w_t L^2}{17\cdot8}$	$\dfrac{w_t L^2}{10\cdot8}$	British code of practice
4	$\dfrac{w_s}{w_d}=1$	$\dfrac{w_t L^2}{10\cdot9}$	$\dfrac{w_t L^2}{9\cdot5}$	$\dfrac{w_t L^2}{16}$	$\dfrac{w_t L^2}{10\cdot3}$	British code of practice
5	$\dfrac{w_s}{w_d}=2$	$\dfrac{w_t L^2}{10\cdot6}$	$\dfrac{w_t L^2}{9\cdot3}$	$\dfrac{w_t L^2}{14\cdot4}$	$\dfrac{w_t L^2}{9\cdot8}$	British code of practice
6	$\dfrac{w_s}{w_d}=3$	$\dfrac{w_t L^2}{10\cdot4}$	$\dfrac{w_t L^2}{9\cdot2}$	$\dfrac{w_t L^2}{13\cdot7}$	$\dfrac{w_t L^2}{9\cdot6}$	British code of practice
7	w	$\dfrac{w_t L^2}{11}$	$\dfrac{w_t L^2}{10}$	$\dfrac{w_t L^2}{11}$	$\dfrac{w_t L^2}{16}$	American Concrete Institute
8	w_t	$\dfrac{w_t L^2}{11}$	$\dfrac{w L^2}{11}$	$\dfrac{w_t L^2}{16}$	$\dfrac{w_t L^2}{16}$	Russian practice

Figure 7

thus an under-reinforced section is ensured. The moment co-efficients used in Russian practice can be obtained as follows. Referring to *Figure 8*, consider the equilibrium of beam AB in which the plastic hinge develops in the span and at B. This represents the mechanism of failure. For the equilibrium of AA' then $M_u = w_t \dfrac{a^2}{2}$ and for the equilibrium of BB' then

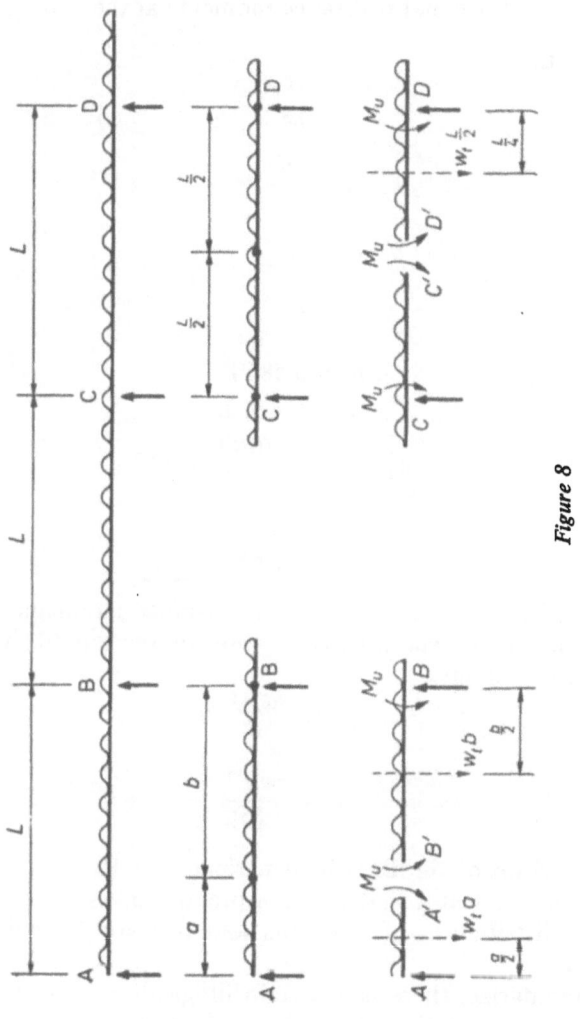

Figure 8

$2M_u = \dfrac{w_t b^2}{2}$ for equal resistance moments at the support and in the span.

thus
$$\tfrac{1}{2} = \frac{a^2}{b^2}$$

$$a = \frac{b}{2^{\frac{1}{2}}}$$

now
$$a + b = L$$

thus
$$\frac{b}{2^{\frac{1}{2}}} + b = L$$

hence
$$b = 0 \cdot 586\,L$$

$$a = 0 \cdot 414\,L$$

$$M_u = w_t \cdot \frac{a^2}{2} = w_t \frac{(0 \cdot 414)^2}{2} \cdot L^2$$

$$= \frac{w_t L^2}{11 \cdot 65} \simeq \underline{\frac{w_t L^2}{11}}$$

For an interior span with equal resistance moments at the support an in the span, consideration of the equilibrium of CC′ or DD′ will give

$$2M_u = \frac{w_t L^2}{8}$$

thus
$$M_u = \frac{w_t L^2}{16}$$

Comparison of the British, American and Russian recommendations indicates that Russian practice gives a more economical distribution of moments and inelastic behaviour is assumed.

To summarize, there is a certain illogicality in sizing reinforced concrete sections by ultimate load methods while the moments for which those sections have been sized have been found by elastic analysis. Providing the section has adequate plasticity, then moment redistribution can be assumed and a plastic analysis carried out. An approximate evaluation of elastic moments is required to check serviceability at working load. A final consideration is deformation of the members at working load and this is discussed in Chapter 5. It will be shown in the next chapter that British, American and Russian recommendations for the sizing of under-reinforced sections by ultimate load methods give similar results, the under-reinforced section forming the basis of a plastic design.

2

THE BEHAVIOUR OF
REINFORCED CONCRETE
UNDER LOAD

Before considering the behaviour of reinforced concrete members under gradually increasing load, it is necessary to summarize British, American and continental practice regarding the resistance of concrete to various types of force action. This chapter is primarily concerned with bending effects and shear, torsion and bond are covered in subsequent chapters. However, it must be emphasized again that in order to take advantage of moment redistribution at ultimate conditions, sections must be proportioned to preclude the possibility of shear, torsion or bond failure, which are of brittle nature.

THE COMPRESSIVE STRENGTH
OF CONCRETE

For design purposes the strength of concrete in compression is normally related to the 28 day strength of cylinders or cubes. When a compressive load is applied to a concrete element it will undergo longitudinal and lateral strains. The presence of pores in the element will produce tensile stresses at right angles to the compressive stress *(Figure 9a)*, causing failure due to the inability of concrete to resist high tensile stresses. The pores in the concrete will produce stress concentrations with local increase in stress thus indicating the necessity for adequate compaction of the concrete to minimize the number of voids. When a concrete cube is tested *(Figure 9b)* the frictional forces between the steel plates and the concrete faces will modify the direction of lateral deformation. This effect will be increasingly reduced towards the centre of the cube. If the

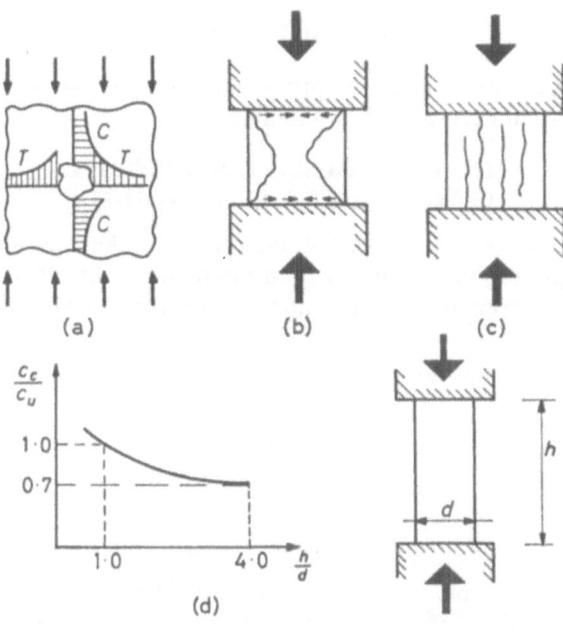

Figure 9

bearing surfaces were frictionless then failure would occur at a lower load and the cracks would be vertical *(Figure 9c)*. The frictional effects on standard cylinders are considerably less than on cubes due to their increased height, resulting in failure at a lower load. The ratio of height to width of a cylinder will obviously affect its failure load and *Figure 9d* indicates the relationship between the height to diameter of the cylinder and the ratio of cylinder to cube strength. Common cube and cylinder sizes for testing purposes are 6in. by 6in. (15 cm by 15 cm) and 12 in. by 6 in. (30 cm by 15 cm) respectively.

The relationship between the cylinder strength C_c and the cube strength C_u is generally in the range

$$C_c = (0 \cdot 7 – 0 \cdot 8)C_u \qquad (2.1)$$

The lower value is more appropriate to higher strength concretes, say in excess of $C_u = 4,500$ lb./in.².

THE COMPRESSIVE STRENGTH
OF CONCRETE IN BENDING

If the stress-strain curve for concrete is taken as approximately parabolic *(Figure 2)* then the form of the stress distribution between the neutral surface and the extreme fibre of

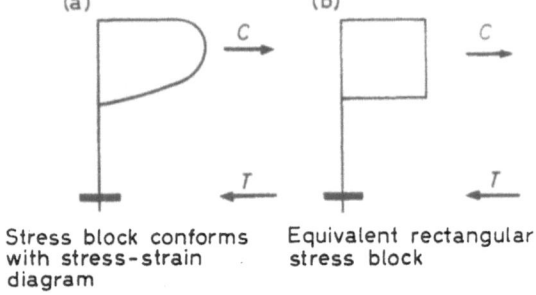

(a)   (b)

Stress block conforms   Equivalent rectangular
with stress-strain   stress block
diagram

Figure 10

the compression zone will also be of parabolic form *(Figure 10a)*. The distribution of stress in the compression zone has been the subject of much research work[2, 6, 7] and many proposals have been made regarding the value of the resultant compressive force and its line of action.

For practical purposes the use of an equivalent rectangular stress block is generally accepted as being adequate *(Figure 10b)*.

THE STRENGTH OF CONCRETE IN TENSION

The strength of concrete in tension can be related to a direct tensile, flexural or cylinder splitting test.

The cylinder splitting test appears to give less variation in results. A compression load is applied along two diametrically

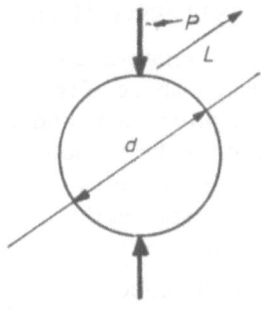

Figure 11

opposed generating lines on cylinders 30 cm long and 15 cm diameter *(Figure 11)*. The tensile strength t is then calculated from the formula

$$t = \frac{2P}{\pi dL}$$

where P is the compression load at splitting failure, d the diameter and L the length of the cylinder.

Tests carried out by the Cement and Concrete Association[26] have shown that the splitting tensile strength of concrete can be represented by the formula

$$t = a \cdot C_u^b$$

where $b = \frac{2}{3}$

$$a = \begin{cases} 1 \cdot 2 \text{ for gravel aggregate} \\ 1 \cdot 3 \text{ for crushed rock aggregate.} \end{cases}$$

The ratio of the splitting tensile strength to the flexural tensile strength of concrete is in the order of 0·5–0·6.

In America the expression $t = 6(C_c)^{\frac{1}{2}}$ is used for calculating splitting tensile strengths. A Russian recommendation is that $t = 0·5(C_u^2)^{\frac{1}{3}}$.

THE SHEAR STRENGTH OF CONCRETE

The shear strength of reinforced concrete is governed by several factors such as the shape of the section; the percentage and distribution of reinforcement and the bending and torsional moments, if any, at the section considered. The combination of shear, bending and torsional stresses may produce high principal tensile stresses and this is discussed in Chapter 6. For bending and torsional shear the total shear stress is limited in the Australian building code to $4(C_c)^{\frac{1}{2}}$ irrespective of the amount of shear reinforcement. This value is some what less than the splitting tensile strength of concrete. A Russian recommendation for the shear strength of concrete is

$$t_{sh} = 0·7(C_c t)^{\frac{1}{2}}$$

THE BOND STRENGTH OF CONCRETE

With smooth reinforcing bars the possibility of movement of the reinforcement relative to concrete is very real. In an ultimate load design bond slip may occur at a lower load than that calculated for bending failure and thus a check must be made on the ultimate bond strength of the concrete. The use of deformed bar reinforcement results in increased bond strength and reduced crack widths. The ultimate bond strength of concrete t_b can be related to its cylinder strength[6] and for deformed bars the following expression can be used

$$t_b = (9·5 - 11·0)\frac{(C_c)^{\frac{1}{2}}}{D}$$

The A.C.I. building code states that for tension bars with sizes and deformations conforming to the American Society for Testing and Materials specification for deformations of deformed steel bars for concrete reinforcement (ASTM-A-305), the ultimate flexural bond stress should not exceed

$$6 \cdot 7 \frac{(C_o)^{\frac{1}{2}}}{D} \text{ or } 560 \text{ lb./in.}^2 \text{ for top bars}$$

and $\quad 9 \cdot 5 \dfrac{(C_o)^{\frac{1}{2}}}{D}$ or 800 lb./in.² for bars other than top bars

where D = bar diameter.

Top bars are horizontal bars so placed that more than 12 inches of concrete is cast in the member below the bar.

STEEL REINFORCEMENT

Typical stress-strain curves for steel are indicated in *Figure 2*. It can be seen from the curves that the characteristics of mild steel are more suited to the concept of plastic design than those

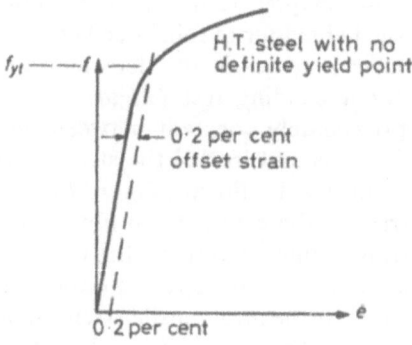

Figure 12

of higher strength steels. However, an idealized stress-strain curve may be assumed for the higher grade steels, and where no definite yield point occurs it may be taken as 0·2 per cent proof stress *(Figure 12)*.

British steels for reinforced concrete work are normally mild steel with yield stress 40,000 lb./in.^{2} or high tensile steel with minimum yield stress 60,000 lb./in.^{2} It is general practice to adopt high tensile deformed bars for reinforced concrete work. The use of deformed bars with the advantages of increased tensile strength and improved crack control outweighs the slight increased cost per ton above that of mild steel.

REINFORCED CONCRETE UNDER LOAD

The following assumptions are generally made in the elastic sizing of a reinforced concrete beam in bending.

(a) The steel and concrete stresses are within the elastic range from zero to working load.

(b) The concrete has negligible resistance to tensile stress.

(c) There is no relative movement of the reinforcement (bond slip) in the concrete.

(d) Plane cross sections of members before stressing remain plane under load.

As indicated in Chapter 1, the above assumptions do not represent the true behaviour of reinforced concrete under load. Consider the various stages in the deformation of a concrete beam loaded for a bending test *(Figure 13a)*. The bending moment is approximately constant between points C and D, and when the beam is first loaded the stresses will be low and the stress diagram will be linear *(Figure 13b)*. The maximum compressive stress in the extreme top fibres will be on the approximately straight line portion of the concrete stress-strain curve. The maximum tensile stress in the extreme bottom fibres will be below the flexural tensile strength of the concrete. As the load increases the tensile strength of the concrete will be reached but the compressive stress will still be within the approximately elastic range. A slight increase in load will cause the concrete to crack at the extreme tension fibres and the tension will be transferred to the steel provided there is sufficient area of steel to take this tension without yield and

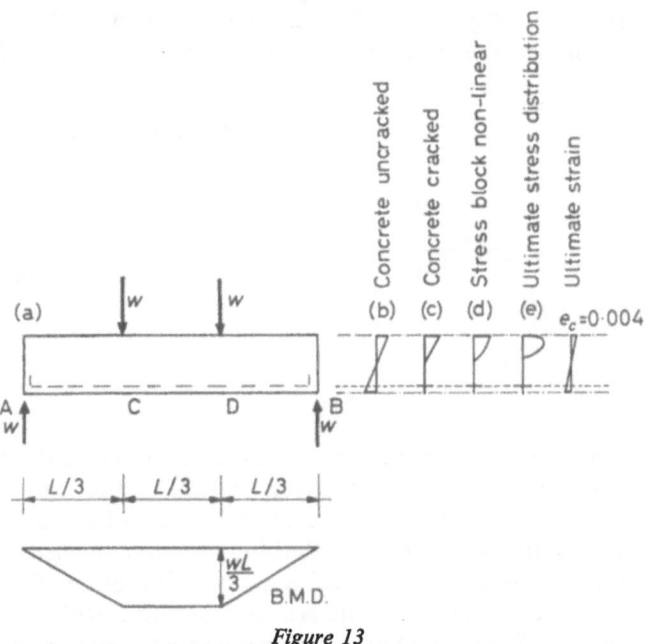

Figure 13

there is adequate bond between the steel and concrete *(Figure 13c)*. With further increase in load the depth of the cracks will increase but the steel bonded to the concrete between the cracks will continue to take the tension. In the compression zone the extreme fibre stresses will no longer be within the straight line portion of the stress-strain curve and the stress diagram will become curved *(Figure 13d)*. If the bond between the steel and the concrete is inadequate the bars will be pulled through the concrete. Inadequate bond results in the formation of much larger cracks and premature failure of the beam will occur. This can be prevented by the use of end anchorage and deformed bars. Deformed bar reinforcement gives a much better crack distribution resulting in a larger number of much smaller cracks *(Figure 14)*. The failure mechanism of the beam will depend on the percentage of reinforcement $p = A_{st}/bd$.

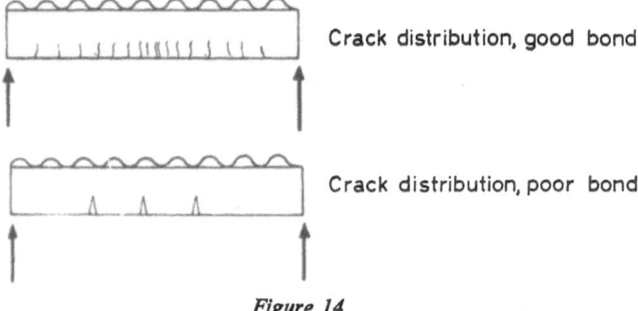

Crack distribution, good bond

Crack distribution, poor bond

Figure 14

If the value of p is low then the steel will yield before the concrete reaches its limiting compressive strain. If the load is increased the cracks will increase in depth resulting finally in crushing of the concrete as its limiting strain is reached *(Figure 13e)*. The limiting strain of concrete without any form of binding varies according to its crushing strength with a range from 0·002 to 0·004, the higher value relating to lower crushing strengths. For normal reinforced concrete work the ultimate strain is taken as 0·003 to 0·004. If the value of p is high then the limiting strain of the concrete will be reached before the steel has reached its yield stress, resulting in a sudden and often explosive failure. As mentioned in Chapter 1 this type of failure is not desirable and for plastic design of concrete sections the percentage of reinforcement is adjusted to ensure that failure is gradual, that is, the section is under-reinforced. Considering the equilibrium of a section at ultimate conditions and assuming an equivalent rectangular stress block, then according to CP 114 (*Figure 15*)

thus
$$C = \frac{4}{9} C_u b \frac{d}{2} = T = A_{st} f_{yt}$$

$$p = \frac{A_{st}}{bd}$$

thus
$$\frac{2}{9} C_u bd = p bd f_{yt}$$

then
$$p = \frac{2}{9} \frac{C_u}{f_{yt}} \qquad (2.2)$$

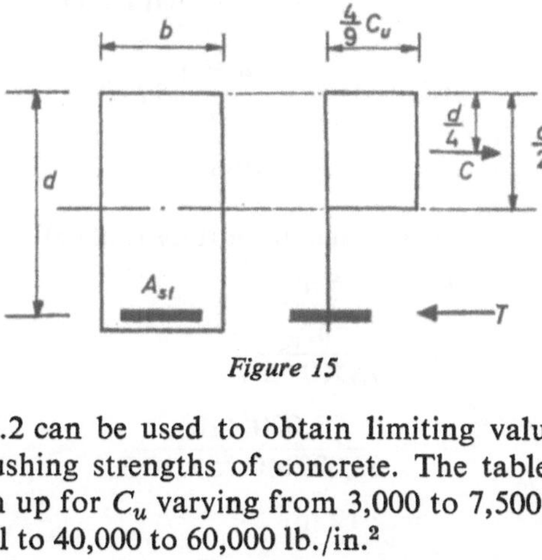

Figure 15

Equation 2.2 can be used to obtain limiting values of p for various crushing strengths of concrete. The table below has been drawn up for C_u varying from 3,000 to 7,500 lb./in.^{2} and for f_{yt} equal to 40,000 to 60,000 lb./in.^{2}

C_u lb./in.^{2}	p for $f_{yt}=60,000$ lb./in.^{2}	p for $f_{yt}=40,000$ lb./in.^{2}
3,000	1·11	1·66
3,750	1·39	2·08
4,500	1·66	2·50
6,000	2·22	3·35
7,500	2·77	4·16

It will be useful at this stage to compare British, American and Russian equations for the ultimate moment of resistance of singly reinforced concrete sections. These equations will be related to a simple example.

Example 2.1

A singly reinforced concrete beam of width 12 inches is to be designed for an ultimate bending moment of 1,000 kip. in., with $C_u = 4,500$ lb./in.^{2} and $f_{yt} = 60,000$ lb./in.^{2}

(a) CP 114 recommendations

$$M_u = \frac{1}{6} C_u \, bd^2 \text{ (see Chapter 1)}$$

thus

$$1,000 \times 10^3 = \frac{4,500}{6} 12 \, d^2$$

$$\therefore \; d = \underline{10 \cdot 5} \text{ in.}$$

As the compression zone has been fully utilized

$$l_a = 0 \cdot 75 \, d$$

hence

$$A_{st} = \frac{M_u}{f_{st} \times 0 \cdot 75 \, d}$$

$$= \frac{1,000 \times 10^3}{60,000 \times 0 \cdot 75 \times 10 \cdot 5}$$

$$= \underline{2 \cdot 12} \text{ in.}^2$$

(b) A.C.I. Building Code Requirements

According to the American Concrete Institute building code requirements for reinforced concrete (A.C.I. 318-63) section 1601, the ultimate design resisting moment of singly reinforced beams is

$$M_u = \phi \left[A_{st} f_{yt} \left(d - \frac{a}{2} \right) \right] \tag{2.3}$$

The coefficient ϕ provides for the possibility of variations in material strengths, workmanship, dimensions, quality control etc. and is taken as 0·9 for flexure,

$$a = \frac{A_{st}f_{yt}}{0\cdot85C_cb}$$

It is also stated that the reinforcement ratio $p = A_{st}/bd$ should not exceed 0·75 of the value that produces a balanced

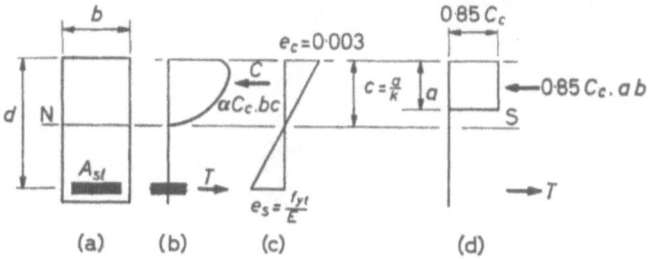

Figure 16

condition *(Figure 16)*, that is the concrete reaches its limiting strain of 0·003 at the same time as the steel reaches its yield stress. Using the equivalent stress block it should be noted that the depth a is related to the true depth c to the neutral surface by a factor k, which is taken as 0·85 for C_c up to 4,000 lb./in.^{2} and reduced by 0·05 per 1,000 lb./in.^{2} for C_c in excess of 4,000 lb./in.^{2} (In the A.C.I. code C_c is represented by f_c'). The following relationship obtains from *Figure 16c*.

$$\frac{e_c}{e_c+e_s} = \frac{c}{d}$$

$$c = \frac{e_c}{e_c+e_s}\,d$$

$$= \frac{0\cdot003}{\dfrac{f_{yt}}{E_s}+0\cdot003}\,d \quad \begin{array}{l}\text{for balanced}\\\text{condition}\end{array}$$

now for
$$C = T$$
$$\alpha C_c b c = A_{st} f_{yt}$$
$$c = \frac{A_{st} f_{yt}}{\alpha C_c b}$$

Where α is a parameter and relates the average stress to C_c.

The steel percentage
$$p = \frac{A_{st}}{bd}$$

This gives
$$c = \frac{p}{\alpha} \frac{f_{yt}}{C_c} d$$
$$= \frac{0 \cdot 003}{\dfrac{f_{yt}}{E_s} + 0 \cdot 003} d$$

For E_s 29×10^6 lb./in.^{2} and $\alpha = 0 \cdot 85\, k$

$$p = \frac{0 \cdot 85 k C_c}{f_{yt}} \frac{87{,}000}{87{,}000 + f_{yt}} = p_b$$

The above expression for p is that at balanced conditions and is termed p_b. In order to ensure an under-reinforced section the maximum value of p is limited to $0 \cdot 75\, p_b$. Applying the A.C.I. formula 2 to Example 2.1, then for $C_c = 0 \cdot 7\, C_u$

$$C_c = 0 \cdot 7 \times 4{,}500 = 3{,}150 \text{ lb./in.}^2$$

$$a = \frac{A_{st} f_{st}}{0 \cdot 85 C_c b} = \frac{2 \cdot 12 \times 60{,}000}{0 \cdot 85 \times 3{,}150 \times 12} = 3 \cdot 96 \text{ in.}$$

$$M_u = \phi \left[A_{st} f_{yt} \left(d - \frac{a}{2} \right) \right]$$

$$= 0 \cdot 9 \times 2 \cdot 12 \times 60{,}000 \,(10 \cdot 5 - 0 \cdot 5 \times 3 \cdot 96)$$

$$= 975{,}000 \text{ lb. in.}$$

now
$$p_b = \frac{0.85 \times k \times C_c}{f_{yt}} \frac{87,000}{87,000 + f_{yt}}$$

$$k = 0.85 \text{ for } C_c = 3,150$$

$$p_b = \frac{0.85 \times 0.85 \times 3,150}{60,000} \times \frac{87,000}{147,000} = 0.0225$$

For $A_{st} = 2.12$ in.^{2}, $b = 12$ in. and $d = 10.5$ in.

$$p = \frac{2.12}{10.5 \times 12} = 0.0168$$

This is just within the limiting value of 0.0169.

(c) Russian Recommendations

Russian recommendations for the maximum stress in the equivalent stress block give the relationship shown in *Figure 17a*. From *Figure 17b* the depth of the stress block is given by

$$2,500 \times 12\,a = 2.12 \times 60,000$$

hence
$$a = 4.24 \text{ in.}$$

$$M_u = 2.12 \times 600,000\,(10.5 - 2.12)$$
$$= 1,065,000 \text{ lb. in.}$$

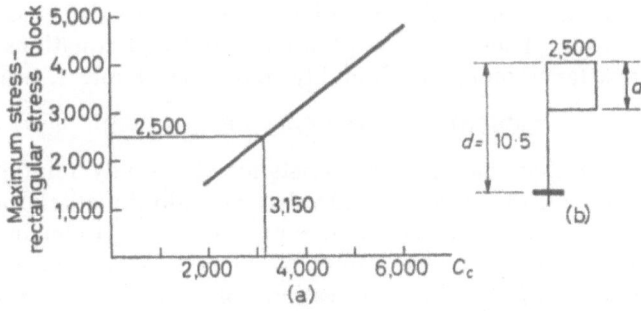

Figure 17

The three results compare reasonably well and the final step is to check the Russian requirements for the under-reinforced condition which is

$$S_c \text{ not greater than } S_0$$

where S_c is the static moment of the concrete area of the compression zone in relation to the centroid of the tensile reinforcement and S_0 is the static moment of the entire effective concrete area in relation to the centroid of the tensile reinforcement

$$S_0 = \frac{bd^2}{2}$$

$$= \frac{12 \times 10 \cdot 5^2}{2} = 662$$

$$S_c = 12 \times 4 \cdot 24 \times 8 \cdot 38 = 427$$

$$\frac{S_c}{S_0} = \frac{427}{662} = 0 \cdot 645$$

thus the under-reinforced condition is satisfied.

These results indicate that the CP 114 recommendations of an average stress of $4/9 \; C_u$ and a limiting depth of $0 \cdot 5 \; d$ for the compression zone approximate to the American and Russian requirements for an under-reinforced section. It is proposed to adopt these parameters for the sizing of reinforced concrete sections throughout the rest of the text due to their ease of application. They can also be used to develop equations for the resistance moment of doubly reinforced T and L beams.

DOUBLY REINFORCED BEAMS

If the ultimate moment of resistance of a singly reinforced section is less than the required value then additional reinforcement will be required in the compression zone. Referring to *Figure 18* let f_{yc} = yield stress of concrete in compression, A_{sc} = area of steel in compression zone, and d_1 = depth to centroid of compression steel,

For equilibrium

$$A_{st}f_{yt} = A_{sc}f_{yc} + \frac{4}{9}\,C_ub\frac{d}{2}$$

thus

$$M_u = A_{sc}f_{yc}\,(d-d_1) + \frac{4}{9}\,C_ub\frac{d}{2}\,\frac{3}{4}\,d$$

$$= A_{sc}f_{yc}\,(d-d_1) + \frac{1}{6}\,C_ubd^2$$

and

$$A_{st} = \frac{2}{9}\,\frac{C_u}{f_{yt}}\,bd + A_{sc}\frac{f_{sc}}{f_{yt}}$$

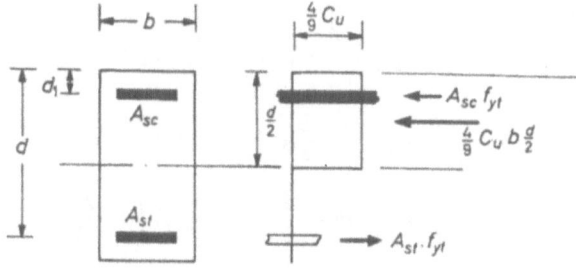

Figure 18

From the above the additional resistance moment required is equated to $A_{sc}f_{yc}\,(d-d_1)$ which gives the required value of A_{sc}. Hence A_{st} can be calculated.

$$f_{yc} = 36,000 \text{ lb./in.}^2 \text{ for mild steel and}$$

$$46,000 \text{ lb./in.}^2 \text{ for high tensile steel.}$$

T and L Beams

The effective width B for T and L beams can be obtained from CP 114 (clause 311e) or from E.C.C. draft regulations, section R4 115. Referring to *Figure 19* let D_s = depth of slab and b_r = width of rib.

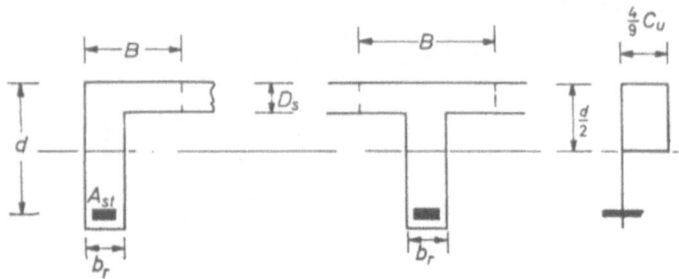

Figure 19

For equilibrium

$$A_{st}f_{yt} = \frac{4}{9}\,C_u b_r \frac{d}{2} + \frac{4}{9}\,C_u(B-b_r)$$

thus $\quad M_u = \frac{4}{9}\,C_u b_r \frac{d}{2}\,\frac{3}{4}\,d + \frac{4}{9}\,C_u(B-b_r)\left(d-\frac{D_s}{2}\right)$

$$= \frac{1}{6}\,C_u b_r\,d^2 + \frac{4}{9}\,C_u(B-b_r)\left(d-\frac{D_s}{2}\right)$$

The above expression assumes full utilization of the concrete compression zone. This is rarely required and if the area of reinforcement is fixed the depth of the concrete compression zone required is obtained by considering the longitudinal equilibrium, that is equating the total tensile force to the total compressive force.

MINIMUM PERCENTAGE
OF REINFORCEMENT

To avoid failure of the steel occurring simultaneously with the cracking of the concrete, the percentage of reinforcement must be such that the total steel force exceeds the tensile resistance of the concrete. Let the depth of the tensile zone from the top fibres be x and the overall depth of the section D *(Figure*

20). Then equating the total steel force to the total tensile force

$$A_{st}f_{yt} = \tfrac{1}{2}t_f b(D-x)$$

Where p_t = percentage of steel relating to tensile zone and
t_s = flexural tensile strength of concrete.

$$p_t = \frac{A_{st}}{b(D-x)}$$

then $\qquad p_t b(D-x)f_{yt} = \tfrac{1}{2}t_f b(D-x)$

$$p_t = \tfrac{1}{2}\frac{t_f}{f_{yt}}$$

If t_f is taken as approximating to $C_u/12$

then $\qquad\qquad\qquad p_t = \dfrac{C_u}{24f_{yt}} \qquad\qquad (2)$

Having considered the behaviour of reinforced concrete sections under load and development of equations for the ultimate resistance moment of reinforced concrete sections, the next stage is to discuss the ultimate load behaviour of prestressed concrete members.

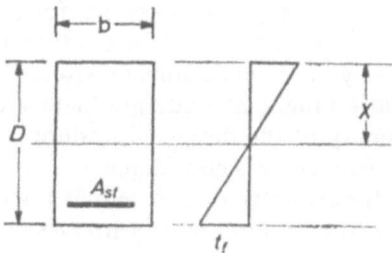

Figure 20

3

THE BEHAVIOUR OF PRESTRESSED CONCRETE UNDER LOAD

Before considering the behaviour of prestressed concrete members between working and ultimate load it is necessary to summarize the elastic method of analysis for simply supported and continuous beams. It is assumed that the reader is aware of the basic methods of prestressing, that is, 'pre-tensioning' and 'post-tensioning'. This chapter will be devoted to flexural considerations, the problems of shear and torsion being considered in Chapter 5.

It is not possible in a book of this size to cover aspects of prestressed work such as stress concentrations at anchorages, losses in prestress force and stressing systems. For detailed information on the above the reader should consult references 8, 9 and 10. As a rough guide, the percentage loss in prestress force due to creep, shrinkage etc. for pre-tensioned and post-tensioned work may be taken as 20 and 30 respectively. The elastic analysis of continuous prestressed concrete beams is not generally taught at undergraduate level thus a fairly detailed summary of the design procedure is included before ultimate load behaviour is considered.

The basic disadvantage of concrete as a structural material is its inability to withstand tensile stresses in excess of about 1/12th of its ultimate strength in compression. The concept of prestressing concrete is to introduce sufficient compressive stress in a member to eliminate tensile stresses induced by self weight, construction and superimposed loads. The development of prestressing can largely be attributed to E. Freyssinet who first conceived the idea in 1904. As the tensile stress in a member due to loading will vary according to the distribution of bending moment, the prestress force can be applied at varying eccentricity to counteract the change in tensile stress along the length of the beam. Referring to *Figure 21* the eccentricity *e*, of the prestressing force P_H can be adjusted

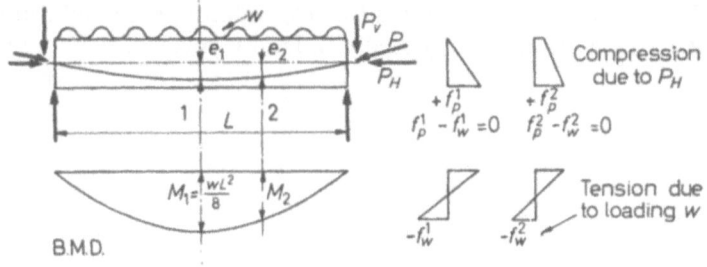

Figure 21

to ensure that at all sections the distribution of stress will be entirely compressive. As the line of action of the prestress force *P* is inclined to the longitudinal of the beam, the vertical component P_V will reduce the shear due to dead and superimposed loading at the section considered. The stress distribution due to the prestress force P_H at any section may be calculated from the expression

$$f_p = \frac{P_H}{A} \pm \frac{P_H e}{Z} \qquad (3.1)$$

where A = area of cross section and Z = section modulus.

Varying eccentricity is normally associated with post-tensioning, the prestressing force being transmitted to the concrete by means of end anchorages. For a uniformly distributed load on a simply supported beam the variation in bending moment will be parabolic for a member of constant depth and thus the prestressing tendon profile will also be parobolic to ensure that the tensile stresses due to loading are exactly counteracted. Another approach to this is the concept of load balancing which was developed by T. Y. Lin[11]. Consider a beam with a parabolic tendon profile *(Figure 22)*. If the ratio e/L is small, say less than 1/12, the radial forces w_p may be taken as vertical. The value of w_p can be adjusted to balance all or part of the applied load w. Considering the equilibrium of a parabolic cable subjected to a load w_p per unit length and

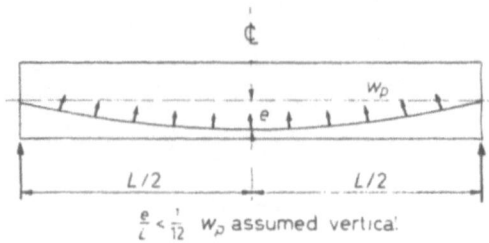

$$\frac{e}{L} < \frac{1}{12} \quad w_p \text{ assumed vertical}$$

Figure 22

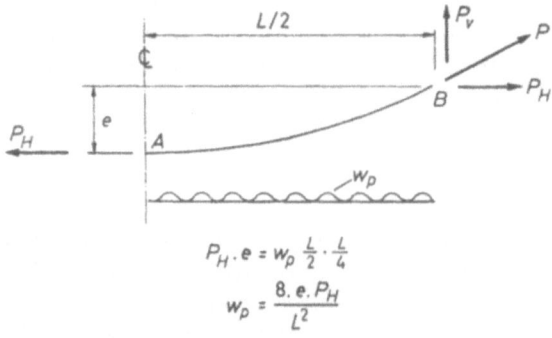

$$P_H \cdot e = w_p \frac{L}{2} \cdot \frac{L}{4}$$

$$w_p = \frac{8 \cdot e \cdot P_H}{L^2}$$

Figure 23

resolving vertically *(Figure 23)*

$$w_p \frac{L}{2} = P_V$$

and taking moments about B for equilibrium for one half of the cable as it is symmetrical

$$w_p \frac{L}{2} \frac{L}{4} = P_H e$$

thus
$$w_p = \frac{8eP_H}{L^2}. \qquad (3.2)$$

thus P_H and e can be adjusted to give the required value for w_p. It is not generally possible to balance the total applied load because the dead load only condition may lead to high tensile stresses, due to prestress occuring in the top fibres. This is due to the large eccentricity required to balance dead plus super-imposed load.

Example 3.1

Design a prestressed concrete roof beam to span 50 ft. between simple supports carrying a total superimposed load of 1 kip per foot run. The concrete stresses are limited to 2,250 lb./in.^{2} in compression and 50 lb./in.^{2} in tension.

Assume trial section as in *Figure 24*.

Cross sectional area $A = 30 \times 12 - 6 \times 14$
$= 276$ in.^{2}

Second moment of area $I = \frac{1}{12} \times 12 \times 30^3 - \frac{6}{12} \times 14^3$

$= 27,000 - 1370$

$= 25,630$ in.^{4}

Section modulus $Z = 25,630/15$

$= 1,710$ in.^{3}

Assuming concrete has a density of 150 lb./ft.^{3}

the self weight per foot run $= \dfrac{276}{144} \times 150 = 288$ lb. $= 0.288$ k

Self weight moment mid-span $M_D = 0.288 \times 50^2 \times 1.5 =$
$$= 1,080 \text{ k.in.}$$

Live load moment mid-span $M_L = 1.0 \times 50^2 \times 1.5 = 3,750$ k.in.

For full load balance $w_p = 1.0 + 0.28 = 1.2888$ k/ft.

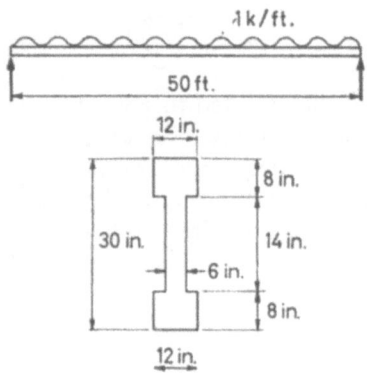

Figure 24

Assuming an eccentricity of $e = 10$ in., which should give adequate cover to the tendons, then

$$P_H = \frac{w_p L^2}{8e} = \frac{1.288 \times 50^2}{8 \times 0.833} = 483 \text{ kips}$$

As the applied load is fully balanced the final stress distribution will be a uniform compression *(Figure 25a)*

$$f = \frac{483 \times 10^3}{276} = 1,750 \text{ lb./in.}^2$$

For self weight only
stress due to self weight $\quad f_{sw} = \dfrac{1,080 \times 10^3}{1,710} = \pm 632 \text{ lb./in.}^2$

Stress due to prestress $\qquad f_p = \dfrac{483 \times 10^3}{276} \pm \dfrac{483 \times 10^3 \times 10}{1,710}$

$$= 1,750 \pm 2,830$$

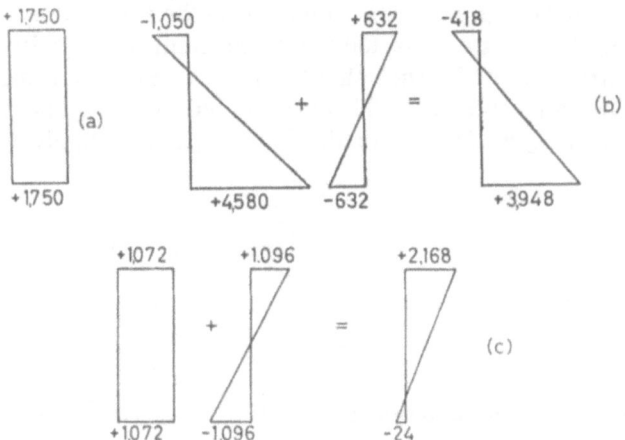

Figure 25

The final stress distribution is shown in *Figure 25b* which is in excess of the stress limits for the concrete. Suppose balance of self weight plus half super load is attempted,

then $\qquad\qquad\qquad w_p = 0.5 + 0.288 = 0.788$

$$P_H = \frac{0.788 \times 50^2}{8 \times 0.833} = 296 \text{ kips.}$$

Direct stress due to $\qquad P_H = \dfrac{296 \times 10^3}{276} = 1,072 \text{ lb./in.}^2$

This acts in conjunction with stress due to unbalanced load of 0·5 kips per foot which gives

$$f = \frac{0.5 \times 3,750 \times 10^3}{1,710} = 1,096 \text{ lb./in.}^2$$

The final stress distribution is given in *Figure 25c*. These values are within the stress limits stipulated.

The concept of load balancing may be applied equally well to continuous beams[11].

For short span members economy can be achieved by maintaining the prestressing tendon at constant eccentricity. The eccentricity must be such that tensile stresses do not occur at any section in the beam. This is governed by the support conditions *(Figure 26)* where the self weight and applied load

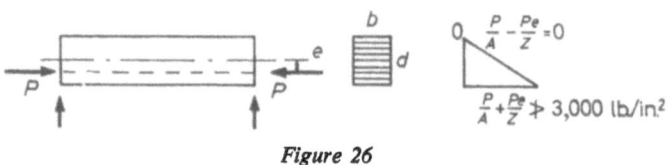

Figure 26

moments are zero and thus the limiting eccentricity is given by

$$\frac{P}{A} - \frac{Pe}{Z} = 0$$

For a rectangular section $Z = \frac{bd^2}{6}$ *(Figure 26)*

hence
$$\frac{P}{bd} = \frac{Pe6}{bd^2}$$

$$e = \frac{d}{6}$$

Constant tendon eccentricity is normally associated with 'pretensioning', the prestressing steel being tensioned against large anchorage blocks before the concrete is placed. The concrete is then placed around the steel and when it is sufficiently mature the prestressing force is transferred to the concrete by cutting the wires at the ends of the members.

With a 'post-tensioned' member the eccentricity can be varied to take the form of the bending moment diagram for self weight and applied loads. The concrete is cast before the steel is tensioned. The tendons are passed through ducts formed in the concrete, stressed by means of jacks and anchored by means of special anchorage devices. The ducts are then grouted under pressure. At the supports the tendon eccentricity can be zero. At mid-span the eccentricity can be greater than that for a 'pre-tensioned' member as the support condition is not

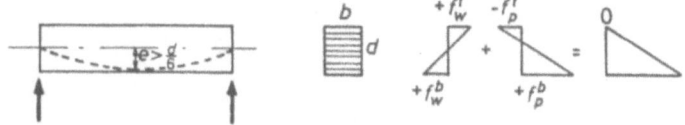

Figure 27

critical. The initial upward camber due to prestress will bring the self weight of the member into play inducing a compressive stress f_w^t in the top fibres. Thus the tensile stress f_p^t due to prestress will be nullified by the stress due to the self weight *(Figure 27)*. The tendon eccentricity will be greater than $d/6$, the limiting value for a rectangular 'pre-tensioned' member.

If a section is subjected to sagging and hogging moments the pre-stress force and eccentricity must be adjusted such that the variation in moment will not induce a tensile stress in the section. This consideration is particularly important in the design of continuous beams.

CONTINUOUS BEAMS

The design of continuous prestressed concrete beams on an elastic basis does not necessarily mean reduction in section size as with reinforced concrete. Further, the analysis is more complex than for reinforced concrete members. Methods of obtaining continuity are illustrated in *Figure 28*. For pre-

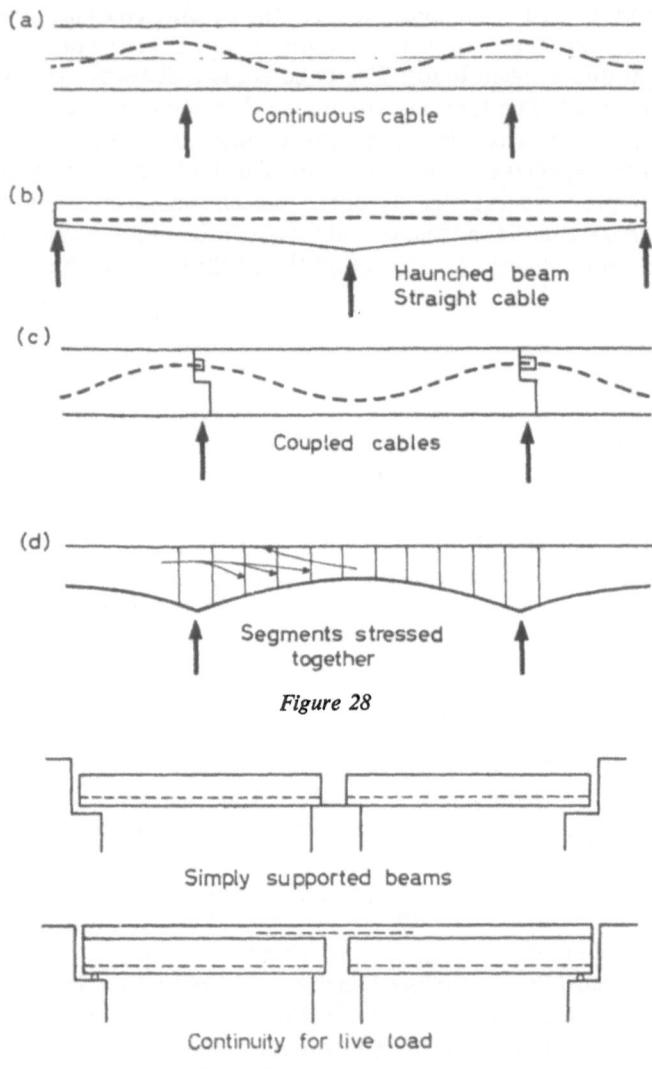

(a) Continuous cable

(b) Haunched beam Straight cable

(c) Coupled cables

(d) Segments stressed together

Figure 28

Simply supported beams

Continuity for live load

Figure 29

tensioned members the use of untensioned steel to obtain continuity *(Figure 29)* for floor systems and bridge decks is quite common and has the advantage of ease of construction. However, secondary moments are set up due to differential shrinkage and restrained creep[9] which may reduce considerably the effective prestress. Continuity is provided for live load only

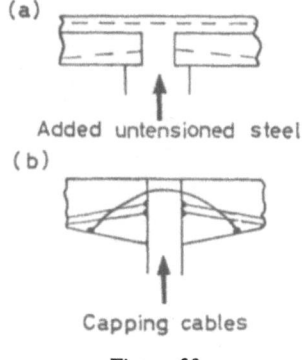

Figure 30

as the precast pre-tensioned beams normally act as permanent shuttering, without propping, for the added *in situ* concrete. Other methods of providing continuity for live load are indicated in *Figure 30*. Continuity for dead and live load may be obtained by the use of continuous cables, the profile being adjusted to suit the variation in bending moment along the beams.

The introduction of an arbitrary cable profile in a beam will, however, introduce secondary moments and reactions. Consider a two-span continuous beam with a tendon profile at a constant eccentricity e *(Figure 31)*. The prestressing force will produce an upward camber and lift the beam off support B. For the beam to rest on the three supports A B and C a downward force R_B is required at B which will produce a deflection equal and opposite to that induced by the prestressing force P.

Equating deflections $\dfrac{Pe(2L)^2}{8EI} = \dfrac{R_B(2L)^3}{48EI}$

$$R_B = \frac{3Pe}{L}$$

$$R_A = R_C = 0.5R_B = 1.5\frac{Pe}{L}$$

Thus moment at B, M_B $\qquad = 1.5\dfrac{Pe}{L}L$

$\qquad\qquad\qquad\qquad\qquad\qquad = 1.5Pe$

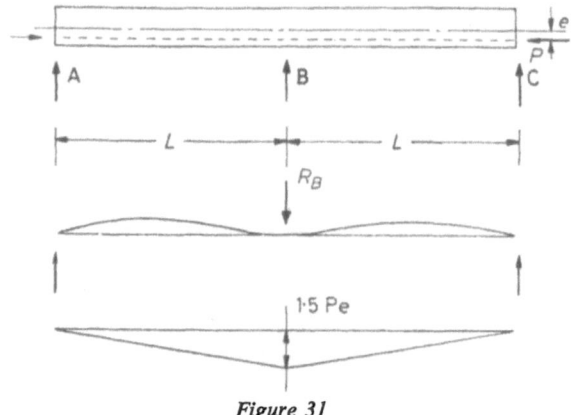

Figure 31

Thus the tendon profile produces a restraint moment at B equal to 1.5 *Pe*. With a more complicated cable profile the above method of equating deflections is not readily applicable and the influence coefficient method of analysis may be adopted.

Example 3.2

A two-span continuous beam *(Figure 32)* is subjected to a total uniformly distributed load of *w* per unit length and the cable profile is parabolic between supports, with zero eccentricity at the three supports and maximum eccentricity e at mid-span. Determine the resultant bending moment at support B.

The structure is made statically determinate by introducing a moment release at B. The moment diagrams due to the prestress and loading w, are m_p and m_w respectively. The moment diagram due to unit moment $x_1 = 1$ applied at B is m_1.

To satisfy the loading conditions

$$f_{11}x_1 + u_1 = 0$$

where*
$$f = \int \frac{m_1^2 d_s}{EI}$$

$$u_1 = \int \frac{m_1 m_p \, d_s}{EI} + \int \frac{m_1 m_w \, d_s}{EI}$$

* Considering flexural energy only the deflection of the released structure in the position c direction of x, is given by

$$u_1 = \frac{m_1 m_p \, d_s}{EI} + \frac{m_1 m_w \, d_s}{EI}$$

The deflection of the released structure in the position and direction of x, for unit value of x, acting alone is

$$f_{11} = \frac{m_1^2 \, d_s}{EI}$$

To satisfy the boundary conditions

$$f_{11}x_1 + u_1 = 0$$

For a detailed explanation of the influence coefficient method of analysis applied to statically indeterminate structures the reader is referred to *Linear Structural Analysis* by P.B. Morice, published by Thames and Hudson.

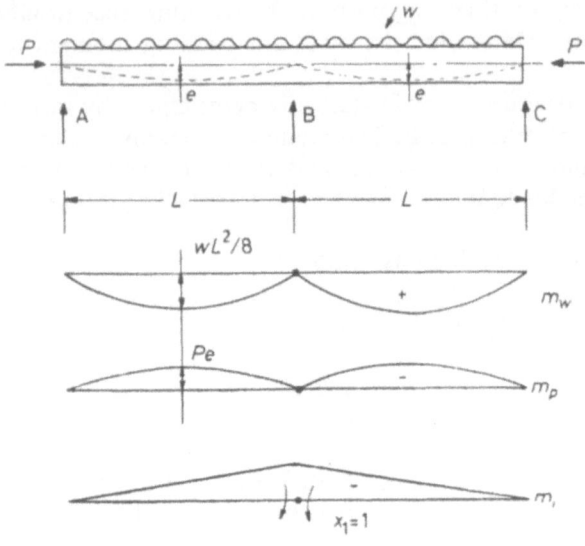

Figure 32

These product integrals may be evaluated using Simpson's rule thus

$$\int m_1^2 \, d_s = \frac{L}{3} 2$$

$$\int m_1 m_p \, d_s = \frac{L}{6} 4 \, pe \, \frac{1}{2} 2$$

$$= \frac{peL}{3} 2$$

$$\int m_1 m_w \, d_s = \frac{L}{6} 4 \frac{wL^2}{8} \frac{1}{2} 2$$

$$= -\frac{wL^3}{24} 2$$

Thus
$$\frac{x_1 L}{3} + \frac{PeL}{3} - \frac{wL^3}{24} = 0$$

$$x_1 = -Pe + \frac{wL^2}{8}$$

The term $wL^2/8$ represents the support moment in a two-span continuous beam subjected to a uniformly distributed load w and $-Pe$ represents the secondary moment induced by the tendon profile. This problem could also be solved using the concept of load balancing.

The above tendon profile produces a support moment which will modify the stress distribution in the beam. If the tendon profile corresponds to the bending moment diagram for any possible loading on a rigidly supported statically indeterminate beam it is referred to as a concordant profile and a tendon laid to that profile will not produce support reactions providing there is no restraint to longitudinal compression. This may be illustrated by a simple example *(Figure 33)*. For the loading

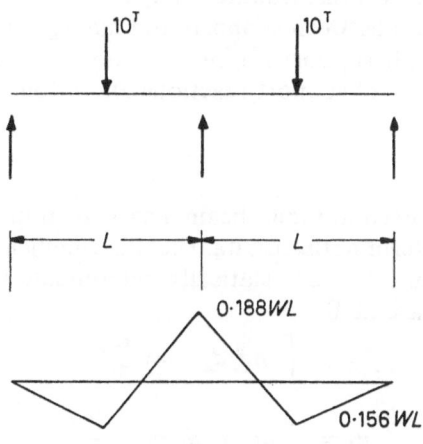

Figure 33

shown the support and mid span moments are 0·188 wL and 0·156 wL respectively.

$$M \text{ support B} = 0·188 \times 10 \times 10 = 18·8 \text{ T. ft.}$$
$$M \text{ span} \qquad = 0·156 \times 10 \times 10 = 15·6 \text{ T. ft.}$$

Let the tendon profile correspond to the moment diagram produced by the point loads $w = 10$ Tons. Using the same procedure as for example 3.2, the product integrals (considering one span only as the moment diagram is symmetrical about B) are

$$f_{11} = 3·33$$
$$u_1 = 0$$

Thus $x_1 = 0$ and hence the tendon profile produces no restraint moment and is concordant.

A concordant tendon profile is not necessarily the most economical as the profile may be such that high friction losses occur due to changes in curvature. The concordant cable profile may be adjusted by the addition or subtraction of a straight line function without altering the distribution of moments. This is termed a linear transformation: any tendon profile consisting of straight lines between supports and having zero eccentricity at an end simple support will produce no bending in the beam but only a series of support reactions and a longitudinal compression.

Example 3.3

A two-span continuous beam has a tendon profile as in *Figure 34*. Determine the resultant bending moment at support B.

The structure is made statically determinate by applying a moment release at B.

$$f_{11} = \int m_1{}^2 \, d_s \quad = \frac{L}{3} 2$$
$$u_1 = \int m_1 m_p \, d_s = -\frac{PeL}{3} 2$$

thus

$$x_1 \cdot \frac{L}{3} = \frac{PeL}{3}$$

$$x_1 = Pe$$

Thus the net bending moment at B is zero. The reaction at B is $2\,P \cos \alpha$.

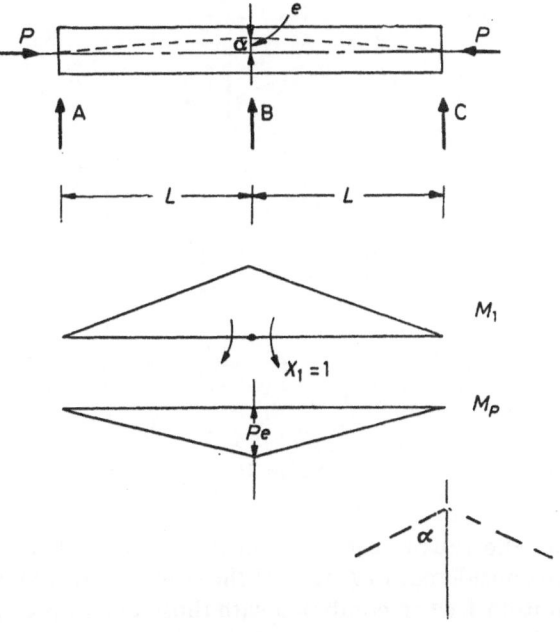

Figure 34

The elastic design of continuous prestressed concrete beams involves the determination of a tendon profile which satisfies the stress conditions for the member with the added consideration that the profile should be concordant if secondary stress is to be eliminated.

THE ADVANTAGES AND DISADVANTAGES
OF CONTINUITY

At any section in a continuous prestressed concrete beam
subjected to varying loading conditions the tendon eccentricity
must be such that the stress distribution due to prestress eli-
minates tensile stresses induced by both hogging and sagging

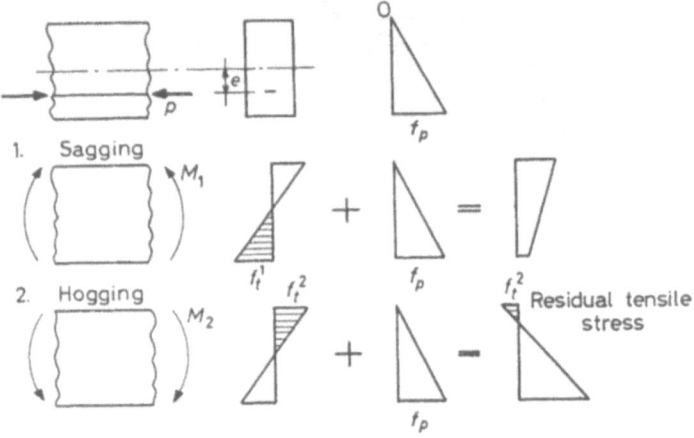

Figure 35

moments and thus at each section the ranges of bending moment are considered. In *Figure 35* the stresses induced by loading condition 1 when combined with those due to prestress are
entirely compressive but loading condition 2 will produce a
residual tensile stress. For a continuous beam with various
loading conditions the moment variation can exceed the simply
supported moment and thus continuity does not necessarily
lead to reduction in construction depth. However, if the dead
load is large some economy in depth can be achieved by continuity. If the dead load is small the beam size will not be
affected because tensile stresses due to dead load can be coun-

teracted by adjusting the tendon eccentricity as indicated previously. With larger spans the additional eccentricity required to offset the load may be such that the tendon would have inadequate cover or even be outside the section. With continuous beams the dead load moments will be less than for the simply supported condition and thus dead load effects will become critical at a larger span than for simply supported beams. It can be seen from the above that if redistribution of moment could be assured at ultimate conditions some economy may be achieved. The possibility of plastic design of continuous prestressed beams will be considered in Chapter 4. The behaviour of a prestressed concrete member between working and ultimate load will now be described.

ULTIMATE STRENGTH
OF PRESTRESSED CONCRETE MEMBERS

The elastic design of prestressed concrete members is generally based on the condition that at working load the stress distribution is entirely compressive. As the load is increased tensile stresses will be induced in the bottom fibres. As the flexural tensile strength of concrete may be several hundred lb./in.^{2}, considerable overload could be incurred before cracks develop. A typical load/deflection curve for a prestressed concrete beam is shown in *Figure 36*, the behaviour being elastic

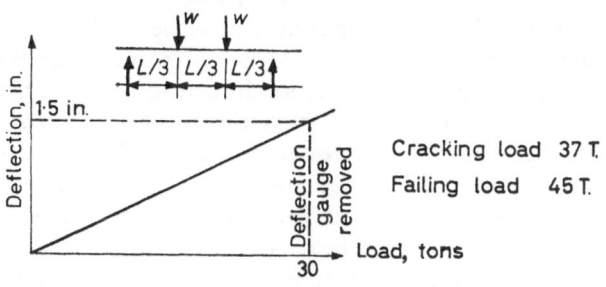

Figure 36

below the cracking load. The behaviour after cracking will be similar to a reinforced concrete beam, there being a departure from the linear stress/strain relationship. In a prestressed concrete beam the steel at working load will have a substantial strain due to the extension required to impart the initial prestress to the concrete. The prestressing steel is generally stressed to about 60 tons/in.² at working load resulting in a strain given by

$$e_{sp} = \frac{f_{st}}{E_s}$$

where f_{st} = stress in steel at working load and E_s = modulus of elasticity of steel.

If $E_s = 29 \times 10^6$ lb./in.² and $f_{st} = 60$ tons/in.² then

$$e_{sp} = \frac{60 \times 2,240}{19 \times 10^6} = 0.004\,6$$

If the bond between the steel and the concrete is effective the load can be increased until the yield stress of the steel is reached or the concrete reaches its limiting strain. A typical stress/strain curve for prestressing steel is shown in *Figure 37*. The characteristics are different from mild steel, there being no definite yield point. If the yield point is taken as 0·2 per cent proof strain, the conditions for an under- or over-reinforced section may be established. For an under-reinforced section the steel will yield before the concrete reaches its limiting strain (unbound) of about 0·003. For an over-reinforced section the concrete will reach its limiting strain before the steel is at the yield stress. It will be seen from the stress/strain curve for prestressing steel that prestressed concrete members will have less plasticity than reinforced concrete members.

An unbonded prestressed beam will behave in a similar manner to a fully bonded beam up to the cracking load. For further increase in load an unbonded member can be considered as a flat tied arch, the increase in load-carrying capacity being dependent on the increase in capacity of the tied arch with increasing deflexion. Typical crack distributions for bonded and unbonded beams approaching ultimate load are indicated in *Figure 38*.

Ultimate load calculations for prestressed concrete members may be carried out by assuming an average stress in the compressive zone at failure of $0 \cdot 85 C_c$ with a limiting concrete

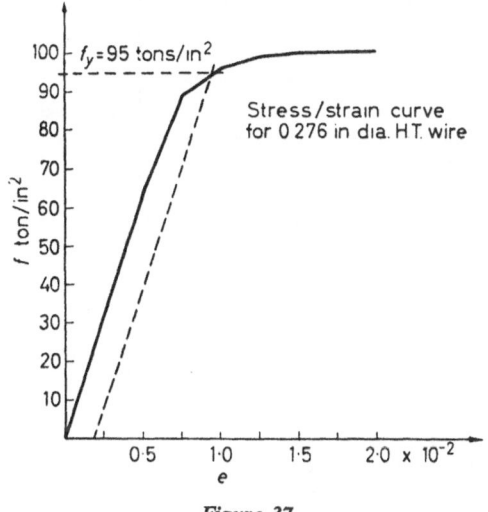

Figure 37

strain of $0 \cdot 003$. From the strain diagram *(Figure 39)* the total strain in the steel can be established. The strain in the concrete at the level of the steel is small and can be ignored. For example if the stress in the concrete is 2,000 lb./in.² at the level of the steel and the modulus of elasticity of the concrete is taken as 5×10^6/lb./in.² then the strain in the concrete is given by

$$e_{cs} = \frac{2,000}{5 \times 10^6} = 0 \cdot 000\ 4$$

The steps in the calculation are as follows. Equating the total compression to the total tension for a rectangular section then

$$A_{st}f_{yt} = 0.85 \; C_c bnd$$
$$= 0.6 \; C_u bnd \text{ for } C_c = 0.7 C_u$$

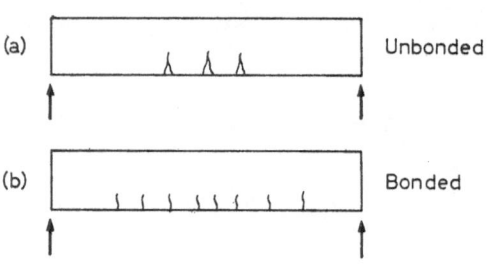

Figure 38

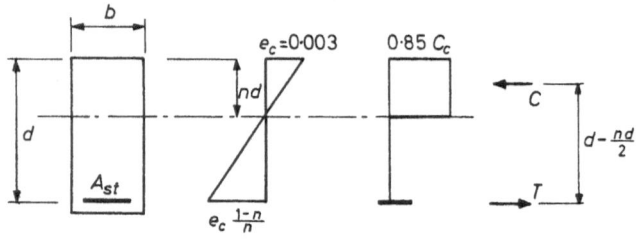

Figure 39

Thus the depth of the neutral surface may be determined. From the strain diagram the total steel strain at failure will be

$$e_{st} = e_{sp} + e_c \frac{d - nd}{nd} \qquad (3.3)$$

If the strain is greater than the assumed strain at yield then the section is under-reinforced. If not the section will be over-reinforced. For an unbonded section the increase in strain between that at prestress and ultimate load is difficult to establish and depends on various factors such as the shape of the bending moment diagram, the number of cracks at failure and the depth of the compression block. Various empirical constants have been suggested which give an expression for the total strain in the form

$$e_{st} = e_{sp} + Kec \frac{d-nd}{nd}$$

where K is an empirical constant which has a range of values 0·1 to 0·25.

Example 3.4

A prestressed concrete beam has an effective depth of 25 in., a width of 24 in. and a steel area of 3·0 in.^{2} *(Figure 40)*. The yield stress of the steel is in accordance with *Figure 37*, and the

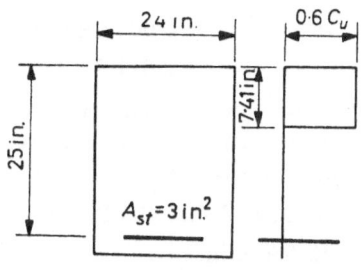

Figure 40

crushing strength of the concrete C_u is 6,000 lb./in.^{2} The steel is stressed to 60 tons/in.^{2} at working load. Determine the ultimate moment of resistance of the section if (a) the steel is fully bonded, and (b) the steel is unbonded with $K = 0·1$.

Equating the total compression to the total tension at ultimate aold

(a) Fully bonded section

$$A_{st}f_{yt} = 0 \cdot 6 C_u bnd$$
$$3 \cdot 0 \times 95 \cdot 2 \times 2,240 = 0 \cdot 6 \times 6,000 \times 24 \times nd$$
$$nd = \frac{3 \cdot 0 \times 95 \cdot 2 \times 2,240}{0 \cdot 6 \times 6,000 \times 24}$$
$$= 7 \cdot 41 \text{ in.}$$
$$n = \frac{7 \cdot 41}{25} = 0 \cdot 296$$
$$\frac{1-n}{n} = \frac{0 \cdot 704}{0 \cdot 296} = 2 \cdot 38$$
$$e_{sp} = \frac{60 \times 2,240}{29 \times 10^6} = 0 \cdot 004 \ 6$$

thus from equation 3.3

$$e_{st} = 0 \cdot 004 \ 6 + 0 \cdot 003 \times 2 \cdot 38$$
$$= 0 \cdot 004 \ 6 + 0 \cdot 007 \ 14$$
$$= 0 \cdot 011 \ 74$$

The strain of $0 \cdot 117 \ 4$ is greater than the $0 \cdot 2$ per cent proof strain and thus the section is under-reinforced. The ultimate moment of resistance is given by

$$M_u = 3 \cdot 0 \times 95 \cdot 2 \times 2,240 \ (25 - 0 \cdot 5 \times 7 \cdot 41)$$
$$= 13,580 \text{ k.in.}$$

(b) Unbonded section

For an unbonded section with $K = 0 \cdot 1$

$$e_{st} = 0 \cdot 004 \ 6 + 0 \cdot 1 \times 0 \cdot 007 \ 14$$
$$= 0 \cdot 005 \ 314$$

This strain is less than the $0 \cdot 2$ per cent proof strain and thus the section is over-reinforced. From *Figure 37*, the stress at a strain of $0 \cdot 005 \ 314$ is $155,000 \text{ lb./in}^2$.

Total tension at this stress $= 155{,}000 \times 3$

$$= 465{,}000 \text{ lb.}$$

equating this to the total compression

$$nd = \frac{465{,}000}{6 \times 6{,}000 \times 24}$$

$$= 5 \cdot 48 \text{ in.}$$

$$n = \frac{5 \cdot 48}{25} = 0 \cdot 216$$

thus

$$\frac{1-n}{n} = \frac{0 \cdot 784}{0 \cdot 216} = 3 \cdot 63$$

$$e_{st} = 0 \cdot 004\ 6 + 0 \cdot 1 \times 3 \cdot 63 \times 0 \cdot 003$$

$$= 0 \cdot 004\ 6 + 0 \cdot 001\ 087$$

$$= 0 \cdot 005\ 687$$

From the stress/strain curve the steel stress is 167,000 lb./in.

Total tension at this stress $= 167{,}000 \times 3$

$$= 501{,}000 \text{ lb.}$$

equating this to the total compression

$$nd = \frac{501{,}000}{0 \cdot 6 \times 6{,}000 \times 24}$$

$$nd = 5 \cdot 8 \text{ in.}$$

$$n = \frac{5 \cdot 8}{24} = 0 \cdot 242$$

thus

$$\frac{1-n}{n} = \frac{0 \cdot 758}{0 \cdot 242} = 3 \cdot 14$$

$$e_{st} = 0 \cdot 004\ 6 + 0 \cdot 1 \times 3 \cdot 14 \times 0 \cdot 003$$

$$= 0 \cdot 005\ 548$$

From the stress/strain curve the steel stress is 162,000 lb./in.²

total tension at this stress $= 162,000 \times 3$
$= 486,000$ lb.

As the change in tensile force is small the ultimate moment of resistance approximates to

$$M_u = 486,000 \ (25 - 0 \cdot 5 \times 5 \cdot 8)$$
$$= 10,740 \text{ k.in.}$$

Thus the ratio of ultimate moments for the bonded and un-bonded condition is

$$= \frac{10,740}{13,580} = 0 \cdot 791$$

A similar procedure can be applied to T and I sections. The British standard code of practice for prestressed concrete BS 115: 1959 gives a method for calculating the ultimate strength of sections having a rectangular compression zone at failure. The above approach is of more general application.

MINIMUM PERCENTAGE OF STEEL

If the tensile force available in the steel over and above the prestressing force is less than the flexural tensile force available in the concrete sudden failure will occur when the concrete cracks. This type of failure can be avoided by specifying a minimum percentage of steel. Referring to *Figure 41* let the

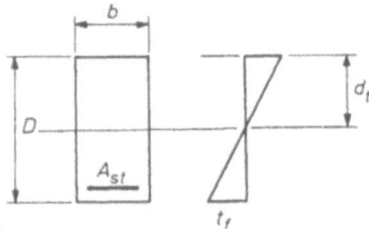

Figure 41

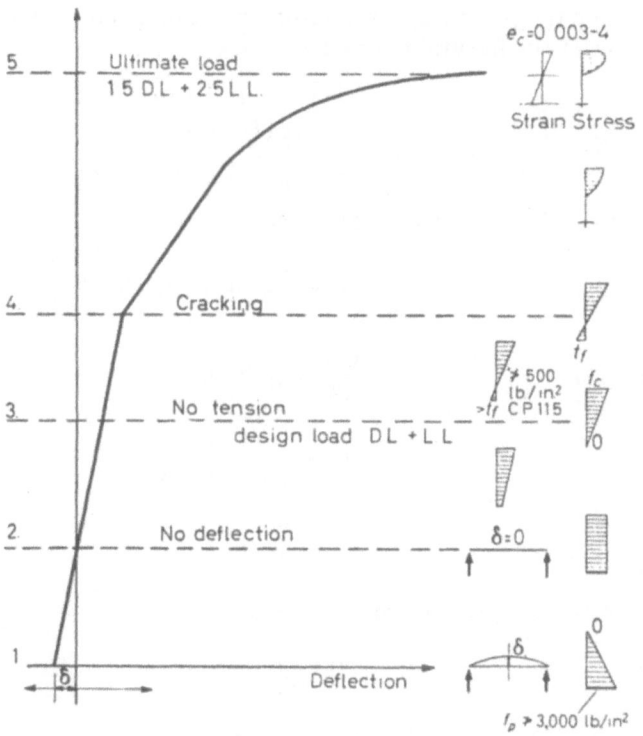

Figure 42

depth of the tensile zone be $D-d_t$ and the flexural tensile strength of the concrete t_f.

If the ultimate stress and the prestressing stress in the steel are f_{yt} and f_p respectively then the tensile force available in the steel above the prestress is

$$T_s = A_{st}(f_{yt}-f_p)$$

The total tensile force available in the concrete is

$$T_c = 0 \cdot 5 \, t_f(D-d_t)b$$

Let p_t be the percentage of reinforcement referred to the cross section of the flexural tensile zone thus

$$p_t = \frac{A_{st}}{b\,(D-d_t)}$$

$$T_s = p_t b (D-d_t)(f_{yt}-f_p)$$

equating T_s to T_c then $0\cdot5t_f = p_t\,(f_{ys}-f_p)$

now if $$t_f = \frac{C_u}{12} \quad \text{and} \quad f_p = 0\cdot6\,f_{yt}$$

then $$\frac{C_u}{24} = p_t 0\cdot4\,f_{yt}$$

$$p_t = \frac{C_u}{f_{yt}\times(24\times0\cdot4)}$$

$$\simeq \frac{C_u}{10}f_{yt}$$

The value of p_t should be in excess of

$$0\cdot1\,\frac{C_u}{f_{yt}}$$

Having outlined the procedure for calculating the ultimate moment of resistance of prestressed concrete sections the next stage is to investigate the possibility of assuming redistribution of moment at ultimate conditions—this is covered in the next chapter. The behaviour of prestressed concrete under load may be summarized by reference to the load/deflection graph, *Figure 42*. Also shown are the stress distributions in the various stages from initial prestress (1), to ultimate load (5). Note that stage (2) represents the full load balance condition.

4

PLASTIC DESIGN OF CONCRETE BEAMS

REINFORCED CONCRETE

The work of J. F. Baker[12] enabled the behaviour to collapse of steel beams and frames to be accurately predicted some twenty five years ago and by the early 1950s the application of plastic theory to the design of continuous beams and portal frames was accepted in the design office. The inelastic behaviour of reinforced concrete was appreciated over fifty years ago by engineers such as E. Freyssinet but until recently there has been little published information regarding the application of a plastic theory to the design of reinforced concrete beams and frames. The 15 per cent moment redistribution allowed in CP 114:1957 makes some concession towards plastic behaviour, but design in accordance with the code is basically an elastic or load factor sizing assuming a substantially elastic distribution of moments. In the late 1940s A. L. L. Baker published several papers on the plastic behaviour of concrete. Research at Imperial College indicated that reinforced concrete was not too brittle for plastic analysis and in 1956 A. L. L. Baker's book[7] was published, and gave recommendations for the plastic hinge design of reinforced concrete beams. The basis of the plastic hinge method of analysis is that a hinge developed in a concrete structure is able to undergo considerable rotation at substantially constant moment. In a continuous reinforced concrete beam plasticity will develop at points of maximum stress for elastic conditions which will generally be at the supports. The steel will yield at the supports over a short length of the beam resulting in a so called plastic hinge. If it is assumed at this stage that the beam has an idealized moment/rotation curve *(Figure 43)* further

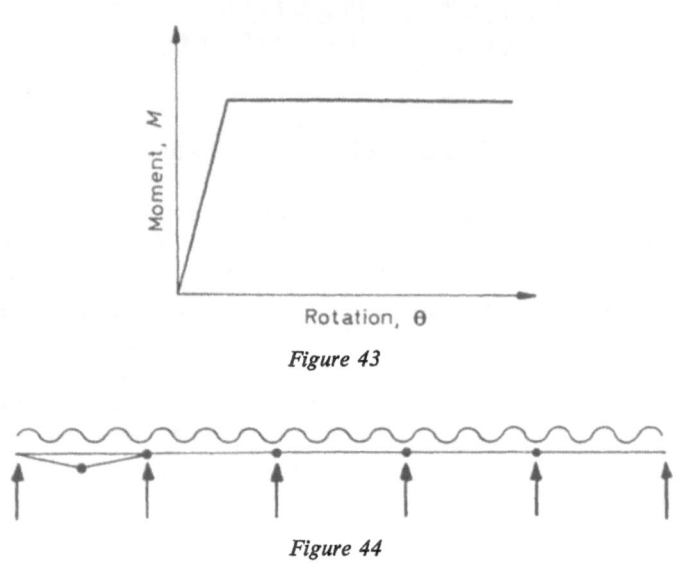

Figure 43

Figure 44

rotation of the hinge could take place at substantially constant moment. For a series of say five continuous beams the structure will become statically determinate if four plastic hinges are developed. The applied load at this condition is termed the ultimate load. The members between the hinges are designed to resist, without the steel yielding, the bending moments incurred when the four hinges have formed. Collapse will occur when a further hinge has formed between the supports *(Figure 44)* resulting in a mechanism. The collapse load will exceed the ultimate load by a small increment of load corresponding to the excess of bending strength provided at midspan to avoid yielding between the supports and thus the formation of a mechanism. This procedure, although simple in concept, requires the calculation of permissible and actual hinge rotations. The actual hinge rotation should be less than the permissible value to ensure that excessive cracking or spalling of the concrete does not take place. Expressions for calculating the permissible hinge rotations are given in A. L. L. Baker's book[7] and also in the report by the Institution of Civil Engineers Research Committee on ultimate load design[2]. These expressions place severe limitations on the allowable rotation of hinges and have subsequently been modified[3]. Calculation of actual hinge rotations may be carried out using the virtual work method or the moment area equation. The above calculations are somewhat tedious and are not generally suitable for the design office. There is also some controversy regarding the necessity to calculate hinge rotations.

Recent research[5] carried out by the Cement and Concrete Association on under-reinforced, balanced, over-reinforced and prestressed sections indicates that brittle failure can be eliminated by the addition of a small quantity of secondary reinforcement. A summary of the results of these tests is as follows:

(1) Under-reinforced concrete beams probably have more than adequate plasticity at failure and should not require any secondary reinforcement at hinges.

(2) The balanced reinforced concrete beams failed in a rather more brittle manner, but very much improved moment/rotation characteristics can be obtained by the use of secondary reinforcement. This took the form of stirrups or helices. The helices appeared to be more economic than stirrups in terms of weight of steel. Baker has also shown that concrete that is well bound[7] has a limiting strain of several times the unbound value of 0·003 to 0·004.

(3) Over-reinforced concrete beams failed in a very brittle manner and to produce an idealized moment curvature relationship a combination of helices and closely spaced stirrups appeared to be necessary.

(4) In beams in which the main reinforcement was a high percentage of high tensile steel, the effect of helical binding was to increase the maximum moment of resistance of the plastic hinge to a value significantly higher than that obtainable without binding.

(5) The rectangular prestressed concrete beam was extremely brittle but again the moment/curvature characteristics were improved by the addition of stirrups and helices.

In order to illustrate the procedure and the limitations imposed by hinge rotation calculations the following example is fully worked out. It is taken from an actual project in which the construction depth was limited in order to obtain maximum space for services. An attempt was also made to keep steel percentages to a reasonable value and not to choke the sections with reinforcement. The calculations were also used as a test case for the general application of the plastic hinge concept in the design office. The results are compared with those obtained from a computer programme developed by A. Weller[13] for the design of continuous concrete beams assuming an elastic distribution of moments with or without a 15 per cent redistribution, section sizing being based on the load factor method. The calculation procedure for the plastic hinge design is summarized below.

(1) Assume plastic hinges occur at supports.
(2) Assume values of support moments.
(3) Calculate section size and reinforcement for bending.
(4) Calculate permissible and actual hinge rotations.
(5) Check serviceability at working load.
(6) Check shear reinforcement for adequate factor of safety.

Stage 1 to 4 are explained more fully in the example and stages 5 and 6 are amplified in subsequent chapters.

Example 4.1

A reinforced concrete beam *(Figure 45)* is continuous over five equal spans of 20 ft. and subjected to a dead load $w_d = 2 \cdot 45$ kips/ft. and a live load $w_s = 1 \cdot 05$ kips/ft. The cube strength of the concrete is 4,500 lb./in.² and the yield strength of the steel 60,000 lb./in.² Design the beam on a plastic hinge basis assuming a load factor of $2 \cdot 0$.

Total load $w_t = w_s + w_d = 2 \cdot 45 + 1 \cdot 05$

$$= 3 \cdot 5 \text{ kips/ft.}$$

Assuming plastic hinges to occur at supports and values of moments as *Figure 45*, then considering the end span and

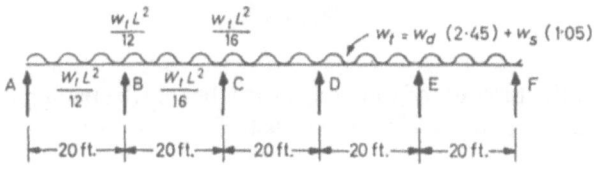

Figure 45

penultimate support moments, the ultimate moment of resistance required will be:

$$M_u = 2 \times 3 \cdot 5 \times 20^2 \times \frac{12}{12}$$

$$= 2,800 \text{ k.in.}$$

Assuming a rib width of 12 in. the resistance moment at the support

B is $\qquad = \dfrac{C_u}{6} bd^2$ for an under-reinforced section

Thus
$$2{,}800 \times 10^3 = \frac{4{,}500}{6} \times 12 \, d^2$$

$$d = 17{\cdot}6 \text{ in. (overall depth say } 20 \text{ in.)}$$

area of reinforcement
$$A_{st} = \frac{2{,}800 \times 10^3}{60 \times 10^3 \times 17{\cdot}6 \times 0{\cdot}75}$$

$$= 3{\cdot}54 \text{ in.}^2$$

In span AB a T section is assumed and from CP 114, clause

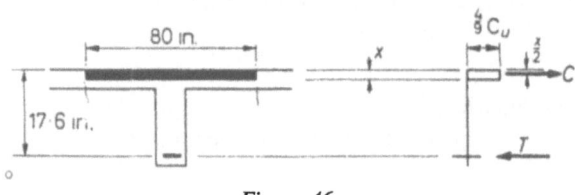

Figure 46

311(e), the breadth of the flange for a beam rib spacing of 16 ft. and a slab depth of 6 in. is the least of

(i) $\dfrac{20}{3} \times 12 \qquad = 80$ in.

(ii) $16 \times 12 \qquad = 192$ in.

(iii) $12 + (12 \times 6) = 84$ in.

Thus for an effective depth of 17·6 in. *(Figure 46)* and flange width of 80 in.

$$\frac{4}{9} \, C_u 80 \, x \left(17{\cdot}6 - \frac{x}{2}\right) = 2{,}800 \times 10^3$$

$$160 \times 10^3 \left(17{\cdot}6x - \frac{x^2}{2}\right) = 2{,}800 \times 10^3$$

Thus $\quad x^2 - 35{\cdot}2x + 35 = 0$

solving the quadratic equation, $x = 1 \cdot 025$ in.

Steel area in span A_{st}

$$= \frac{2,800 \times 10^3}{60 \times 10^3 (17 \cdot 6 - 0 \cdot 501)}$$

$$= 2 \cdot 73 \text{ in.}^2$$

The permissible hinge rotation will first be calculated, assuming the concrete is not bound, in accordance with the

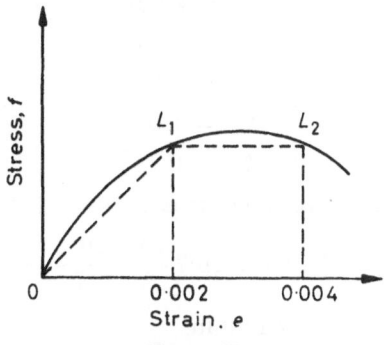

Figure 47

I.C.E. report[1]. If the idealized stress/strain characteristics are as shown in *Figure 47*, L_2 being the limiting strain and L_1 the effective yield point, then the increase in strain between L_1 and L_2 will be at substantially constant moment and the permissible hinge rotation on one side of the critical section is given by

$$\theta_p = \frac{(e_{cu} - e_{ce}) l_p}{n_u d} \qquad (4.1)$$

where e_{cu} and e_{ce} are the strains corresponding to L_2 and L_1. l_p is an equivalent plastic length over which plastic rotation occurs with constant curvature and can be expressed by

$$l_p = k_1 k_2 k_3 \left(\frac{z}{d} \right)^{\frac{1}{4}} d \qquad (4.2)$$

where k_1 = parameter for influence of type of steel, k_2 = parameter for influence of axial load = 1 in this case, k_3 = parameter for influence of grade of concrete, z = distance from initial section to point of contraflexure, approximately, 0·15 span, d = effective depth of section, and n_u = neutral axis factor at failure state L_2.

For this example it is assumed that[1]

$$e_{cu} = 0·004$$
$$e_{ce} = 0·002$$
$$k_1 = 0·9$$
$$k_2 = 1·0$$
$$k_3 = 0·7$$

Substituting the above values in equation 4.2 then

$$l_p = 0·9 \times 0·7 \frac{(0·1520 \times 12)^{0·25}}{17·6} 17·6$$

$$= 13·3 \text{ in.}$$

and the value of n_u may be determined by equating C_u to T_u. Assuming an average stress of 0·85 C_u at failure and $C_c = 0·7 C_u$

Average stress $= 4,500 \times 0·7 \times 0·85$

                       $= 2,680 \text{ lb./in.}^2$

thus $2,680 \times 12\, n_u d = 60,000\, A_{st}$

At support B, $A_{st} = 3·53 \text{ in.}^2$

thus $n_u = \dfrac{60,000 \times 3·53}{2,680 \times 12 \times 17·6}$

                 $= 0·375$

Change in strain between $L_2 L_1$ is $e_{cu} - e_{ce} = 0·004 - 0·002 = 0·002$. Substituting in equation 4.1 then

$$\theta_p = \frac{0·002 \times 13·3}{0·375 \times 17·6} = 0·004$$

This is the rotation on one side of the critical section thus the permissible rotation = $2 \times 0.004 = 0.008$

The actual hinge rotation may be obtained by loading the beam with the M/EI diagram and determining the reactions, or the virtual work method.

Virtual Work Method

In general terms the virtual work equation is

$$\int \frac{M_i \cdot M_o}{EI} \, d_s + X_i \int \frac{M_i \cdot M_i}{EI} \, d_s + \sum X_k \int \frac{M_i M_k}{EI} \, d_s + \theta_i = 0 \quad (4.3)$$

where θ_i = inelastic rotation of hinge, X_i = unknown moment acting at hinge i, M_i = moment due to $X_i = 1$ acting at hinge i, M_k = moment due to $X_k = 1$ acting at hinge k, and M_o is moment due to external load.

For this example $M_o = \dfrac{w_t L^2}{8} = 4{,}200$ k.in. *(Figure 48)*

$$M_i = \frac{w_t L^2}{12} = 0.67 \, M_o$$

$$M_k = \frac{w_t L^2}{16} = 0.5 \, M_o$$

hence $\quad \displaystyle\int \frac{M_i M_o}{EI} \cdot d_s + X_1 \int \frac{M_i^2}{EI} \, d_s + X_2 \int \frac{M_i M_2}{EI} \, d_s$

$$+ X_3 \int \frac{M_1 M_3}{EI} \, d_s + X_4 \int \frac{M_1 M}{EI} \, d_s + \theta_1 = 0 \quad (4.4)$$

Using Simpson's rule for product integrals and substituting in equation 4.4 then

$$-\frac{2}{3} \cdot \frac{M_o L}{EI} + 0.67 \frac{M_o L}{EI} + 0.5 \frac{M_o L}{EI} = \theta_1$$

thus $\qquad\qquad \theta_1 = 0.139 \dfrac{M_o L}{EI}$

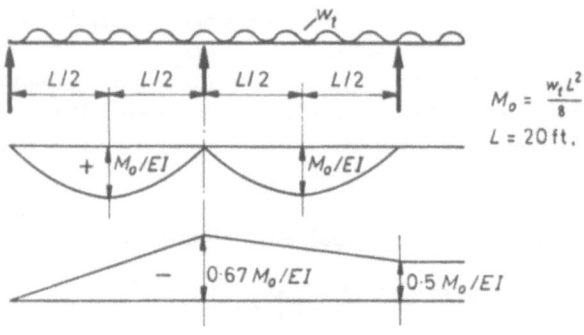

Figure 48

Mohr's Method

Loading the beams with the M/EI diagram and determining the reactions *(Figure 48)* then

$$\theta = \frac{1}{EI}\left(\frac{2}{3}\cdot M_oL\,\frac{1}{2}\cdot 2 - \frac{1}{2}\cdot\frac{2}{3}\cdot M_oL\cdot\frac{2}{3} - \frac{M_o}{2}\cdot\frac{L}{2}\cdot\frac{L}{2}\right.$$

$$\left. - \frac{M_o}{6}L\cdot\frac{1}{2}\cdot\frac{2}{3}\right) = 0\cdot139\,\frac{M_oL}{EI}$$

The final stage is to determine the value of EI. It is assumed that EI is constant[1] between hinges and equal to the value at the idealized elastic limit L_1. For cracked rectangular sections the I.C.E. report gives

$$EI = \frac{\alpha C}{C_u}\frac{1}{e_c}\,(n_1^2 - \gamma n_1^3)bd^3C_u \qquad (4.5)$$

where e_c = strain in concrete at edge of section = $0\cdot002$, n_1d = depth of neutral surface at L_1, c = concrete stress at edge of section, γn_1d = depth to centre of compression *(Figure 49)*, αc = average concrete compressive stress.

The following values are assumed

$$L = 0\cdot67$$
$$\gamma = 0\cdot4$$
$$C = 0\cdot85\ C_u$$
$$n_1 = 0\cdot5$$

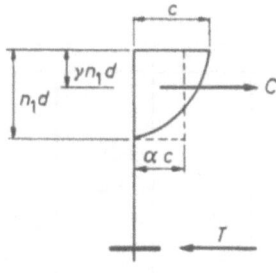

Figure 49

Substituting these values in equation 4.5 then

$$EI = 57\ bd^3\ C_u$$

hence actual hinge rotation

$$= \frac{0\cdot139\ M_oL}{57\cdot bd^3 C_u}$$

$$= \frac{0\cdot139 \times 4{,}200 \times 10^3 \times 20 \times 12}{57 \times 12 \times 17\cdot6 \times 4{,}500}$$

$$= 0\cdot008\ 3$$

Thus the actual hinge rotation approximates to the permissible value at support B. A further calculation can be made for support C and it will be found that the permissible rotation is considerably greater than the actual rotation.

From the above it will be seen that hinge rotation calculations are rather tedious and involve a considerable number of assumptions. A computer programme would minimize the arithmetical work. The hinge calculations indicate that for this example the penultimate support and end span moments of $w_tL^2/12$ are the limiting values in terms of permissible rotation. Further checks must be made for serviceability and shear and these are discussed in the following two chapters.

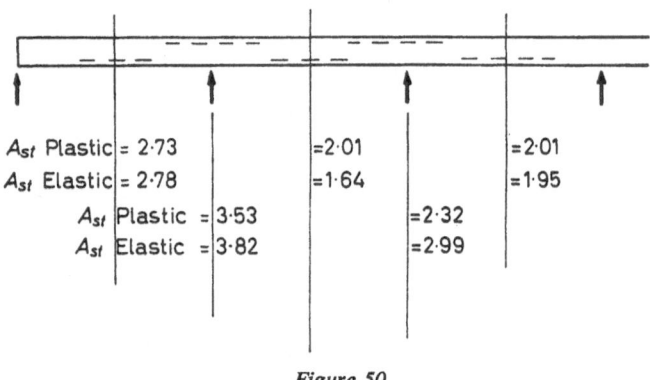

A_{st} Plastic = 2·73   =2·01   =2·01
A_{st} Elastic = 2·78   =1·64   =1·95
          A_{st} Plastic = 3·53   =2·32
          A_{st} Elastic = 3·82   =2·99

Figure 50

A comparison of steel **areas obtained** using the moment coefficients assumed for plastic **analysis** and those obtained from A. Weller's computer programme for elastic analysis with 15 per cent redistribution are indicated in *Figure 50*. The plastic design gives a more favourable distribution of steel. Permissible hinge rotations may be increased considerably by taking into account the binding effect of stirrups as will be shown in example 4.2.

Example 4.2

Using the same design data as for example 4.1 determine the permissible rotation at support B taking into account the binding effect of the stirrups.

The revised formula for calculating θ_p is

$$\theta_p = 0.8(e_{cu} - e_{ce})k_1 k_3 \frac{z}{d} \qquad (4.6)$$

This expression is taken from reference 3 which gives revised formulae for calculating θ_p and e_{ce}. Using the data from example 4.1

$$\theta_p = 0.8(e_{cu} - 0.002) \, 0.9 \times 0.7 \frac{0.15L}{d}$$

The value of e_{cu} is obtained from

$$e_{cu} = 0.001 \, 5(1 + 1.5p'') + (0.7 - 0.1p'') \left(\frac{1}{n_2}\right) \qquad (4.7)$$

Where p'' = percentage of binders (volume ratio), n^2 = ratio of neutral axis depth to effective depth, and c corresponds to n_u in example 4.1 = 0.375.

Suppose the binding consists 4-$\frac{3}{8}$ in. dia bars at 3 in. cross centres, the volume percentage binders will be about 1.5. Substituting in equation 4.7

$$e_{cu} = 0.001 \, 5(1 + 1.5 \times 1.5) + (7 - 1 \times 1.5) \frac{1}{0.375}$$

$$= 0.001 \, 5(1 + 2.25 + 1.46)$$

$$= 0.007 \, 1$$

This is a considerable increase from the limiting value of 0.004 for unbound concrete
Substituting for $e_{cu} = 0.007 \, 1$ in equation 4.6

$$\theta_p = 0.8(0.007 \, 1 - 0.002)0.63 \times 0.15 \frac{20 + 12}{17.6}$$

$$= 0.8 \times 0.005 \, 1 \times 0.63 \times 0.15 \frac{20 + 12}{17.6}$$

$$= 0.005 \, 26$$

This is on one side of the critical section and thus the permissible rotation is

$$= 0.005\ 26 \times 2$$
$$= 0.010\ 52$$

Thus the modified equations 4.6 and 4.7 give a higher permissible rotation as the effect of binders is taken into account.

RECOMMENDATIONS FOR DESIGN

The number of assumptions to be made when calculating permissible and actual hinge rotations, the difficulty of interpreting the information given in references 1 and 3 and the considerable doubt as to the necessity for calculating hinge rotations for under-reinforced beams, has discouraged engineers from using plastic theory for the design of continuous concrete beams.

However, providing the design is based on an under-reinforced condition, the Russian moment coefficients *(Figure 7)* may be adopted without a check being made on hinge rotations and they also give adequate serviceability at working load (see Chapter 6). Another approach would be to increase the percentage moment redistribution from say 15 to 25. Tests by A. H. Mattock[14] on continuous concrete beams with mild and work hardened square twisted steel reinforcement led to the conclusion that redistribution of bending moments by amounts up to 25 per cent did not appear to affect adversely the performance at working or ultimate load. If a penultimate support moment of less than $w_t L^2/12$ is adopted for beams of equal span, then great care must be exercised in checking serviceability and hinge rotations. For other beams the restriction of 25 per cent redistribution should be applied.

PRESTRESSED CONCRETE

If it can be established that full redistribution takes place at ultimate conditions in reinforced or prestressed concrete continuous beams, then the ultimate load may be determined from the calculated values of the ultimate moment at the critical sections and the known self weight moments. If the ultimate moments calculated for the support and span in a continuous beam are M_{u_1} and M_{u_2} respectively *(Figure 51)* then the ultimate load w_u is obtained from

$$\frac{w_u L}{4} + M_D = M_{u_2} + \frac{1}{2} M_{u_1}$$

Hence

$$w_u = \frac{4}{L} \left(M_{u_2} + \frac{1}{2} M_{u_1} - M_d \right)$$

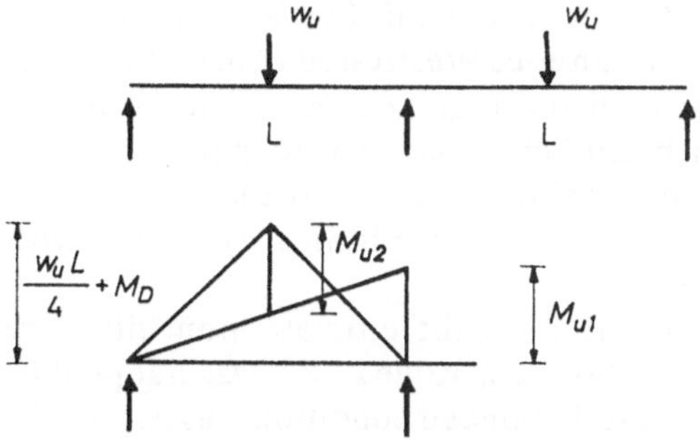

Figure 51

This assumes plastic behaviour, that is, hinges are developed at critical sections and rotation takes place at substantially constant moment without excessive cracking or spalling. From an assumed elastic distribution of moments at working load the tensile stress in prestressed concrete beams may be determined. If this is greater than the flexural tensile strength of the concrete then cracks will form at working load.

Thus for prestressed beams the serviceability requirement at working load could be that the tensile stress is less than the flexural tensile strength of the concrete. However, slight overload may cause cracking which in some circumstances would be undesirable. Thus for both reinforced and prestressed concrete members serviceability checks are essential.

The failure of prestressed concrete beams is, in general, of more brittle nature than for reinforced concrete beams. However, the introduction of secondary reinforcement in the form of stirrups or helices considerably improves the moment/rotation characteristics, as demonstrated by the Cement and Concrete Association Tests[5]. Several series of tests have been carried out to ascertain the amount of redistribution of moment which can occur in continuous prestressed concrete beams. The results are summarized below.

(1) P. B. Morice and H. E. Lewis[15]. The tests were carried out on two continuous prestressed concrete beams. The beams were of constant rectangular cross section 6 in. deep by 4 in. wide, each beam being loaded with point loads at mid-span. The concordant cable profile was made rectilinear for the majority of the beams, but two beams were cast with parabolic cable profiles.

In the calculation of the ultimate moments of resistance at the various sections a k value of 0·9 was adopted (see page 59) thus a near to fully bonded condition was assumed.

Transformations of the maximum possible value were given to the rectilinear cable profile and the test results indicated that large transformations had no significant effect on the ultimate load of the beams. The parabolic cable profile was transformed to a smooth curve along the entire length of the beam and again the transformation had no significant effect on the ultimate load.

The tests indicated that full redistribution of moment did take place and that the ultimate load for a given form of loading depends on the curvature of the cable profile within the span and is independent of the absolute values of the eccentricities.

(2) G. Macchi[15]. Tests were made on three continuous three-span beams of section 10 in. deep and 4 in. wide. For two of

the beams the spans were 6·5 ft., 13·1 ft. and 6·5 ft. and for the third 9·85 ft., 13·1 ft. and 9·85 ft. The beams were tested by a point load at the middle of the centre span and the results indicated that redistribution was incomplete.

(3) Y. Guyon[15]. Tests were made on four two span continuous beams, the spans being 13 ft. 1 in. and the beam section 10 in. deep and 5 in. wide. The beams were loaded equally at mid-span and the results indicated full redistribution took place.

The results of the above and other tests are summarized in a paper by Y. Guyon[15] which also gives a theory for the calculation of the ultimate load of statically indeterminate prestressed concrete structures based on the consideration of plastic hinges, at which the moments are assumed to be known and equal to the resistance moments of these sections.

At the present stage the experimental evidence available does not appear to be sufficient to assume that full redistribution of moment does in fact occur in prestressed concrete continuous beams. A practical method for the possible load-carrying capacity of continuous prestressed concrete beams has been suggested by F. Leonhardt[9]. The procedure is outlined below.

(1) The moment distribution for the required failure load is obtained by elastic analysis. Suppose at a given section the moment coefficients for dead and live load are $w_d L^2/10$ and $w_s L^2/9$, then the total moment at working load is

$$M_w = \frac{w_d L^2}{10} + w_s \frac{L^2}{9}$$

if a factor of safety of $1·5 w_d + 2·5 w_s$ is assumed then the required ultimate moment would be

$$M_u = \frac{1·5 w_d}{10} L^2 + \frac{2·5 w_s}{9} L^2$$

(2) The support moments are now superimposed on the simply supported moment diagrams *(Figure 52)*.

(3) The actual ultimate moment at any particular section is determined from the section size and steel area (Chapter 3).

Suppose at section K the actual ultimate moment is less than the required value. The moment diagram must now be adjusted to ensure that the actual ultimate moment at K (and

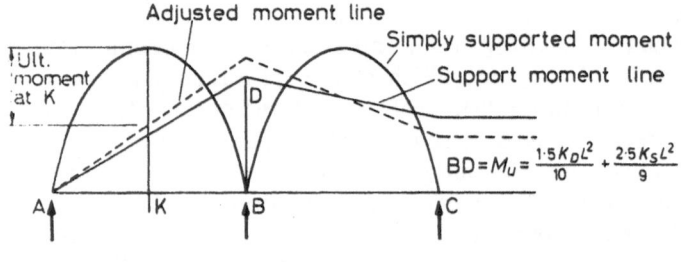

Figure 52

other sections) is not exceeded (broken line). The other sections are checked in a similar manner. Leonhardt[9] gives a table from which can be tentatively estimated the amount of adjustment of the moment diagram that can be made without excessive hinge rotation occuring. This is related to the steel percentage, concrete and steel strengths.

Determination of the ultimate load capacity of continuous prestressed (or reinforced concrete) beams in this manner is very straightforward and could well be adopted as a design office procedure. Further research is required to relate available redistribution of moment to the steel percentage, concrete and steel strengths and the amount of secondary reinforcement.

5

SHEAR AND TORSION

REINFORCED CONCRETE

Shear

The February 1965 amendment to CP 114:1957 does not give recommendations for ultimate load design with regard to shear. Recent tests carried out at the Building Research Station have shown that the permissible shear stress given in Table 6 (CP 114), for beams without shear reinforcement, do not give a load factor consistent with that adopted for bending failure. In order to take advantage of redistribution of moments, the structural members must be proportioned to ensure that failure will occur in bending. The B.R.S. tests have led to recommendations given in the table below for permissible shear stresses in beams without shear reinforcement. The nominal shear stresses in CP 114 are given for comparison, a value of $7/8d$ being assumed for the lever arm.

Cube strength of concrete (lb./in.²)	Recommended limit of nominal shearing stresses Q/bd when stress in longitudinal reinforcement is 20,000 lb./in.² (lb./in.²)	Recommended limit of nominal shearing stresses Q/bd when stress in longitudinal reinforcement is 30,000 lb./in.² (lb./in.²)	Nominal shearing stress Q/bd recommended in CP 114 (lb./in.²)
3,000	50	45	87
4,500	55	50	114
6,000	60	55	140

The above table is taken from a B.R.S. current paper in the Engineering Series No. 6. 'Permissible Shearing Stresses in Reinforced Concrete Beams' by R. Taylor, B.Sc., D.I.C., A.M.I.C.E., which summarizes the results of the tests. The method of loading and the position of the loads on the beam influence considerably its ultimate shear resistance.

If the permissible shear stress is exceeded then according to CP 114 the whole of the shearing force should be resisted by shear reinforcement. Further the shear stress should not exceed four times the permissible value irrespective of the shear reinforcement provided. The B.R.S. recommendations for maximum shearing stresses in beams with shear reinforcement are given below and are again compared with the CP 114 values.

Cube strength of concrete (lb./in.²)	Recommended limit of nominal shearing stresses Q/bd (lb./in.²)	Nominal shearing stress Q/bd recommended in CP 114 (lb./in.²)
3,000	200	350
4,500	300	455
6,000	400	560

A method of proportioning stirrups has also been developed at the Building Research Station and is summarized in Engineering Series current paper No. 8 'A New Method of Proportioning Stirrups in Reinforced Concrete Beams' by R. Taylor, B.Sc., D.I.C., A.M.I.C.E.

The basic difference between this method and the CP 114 procedure is that the new method requires both bending moment and shear force to be considered and is based on the effectiveness of the stirrups in carrying the shear force across an inclined crack. The basic aim of developing the method was to design the shear reinforcement to preclude the possibility of shear failure. There are however several types of shear failure which involve a considerable number of variables such as the shape of the section, the amount and type of longitudinal reinforcement and stopping off points. For a detailed discussion of shear failure see reference 16. The shear crack may be considered to be made up of:

(a) A crack inclined at approximately 45° to the horizontal axis between the level of the longitudinal steel and the level of

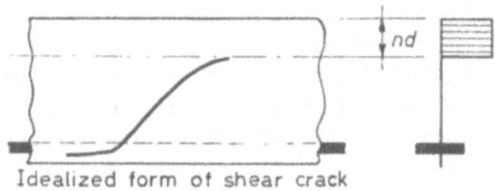

Idealized form of shear crack

Figure 53

the compression zone, that is, the compression zone as it was before some redistribution of stresses occured as a result of the penetration of the crack *(Figure 53)*.

(b) An upward extension of the crack into the original compression zone at an angle somewhat less than 45° to a distance n_d from the top surface, where n_d is the depth of concrete in compression calculated by the load factor method in CP 114.

(c) A backward extension of the crack along the level of the longitudinal tension steel.

The stirrups that are effective in carrying shear forces across an inclined crack comprise not only those stirrups that cross the inclined part of the crack, but owing to the dowel effect of the longitudinal bars, stirrups behind the end of the inclined crack. Vertical stirrups to resist shear should be proportioned according to the equation

$$s = \frac{0 \cdot 9 f_y A_w L}{Q_u} \qquad (5.1)$$

where s = stirrup spacing, A_w = cross sectional area of a stirrup, f_y = yield stress of stirrup steel, Q_u = shear force at ultimate load, and L = length of horizontal projection of effective shear crack.

$$L = K(d_b - d_n) \quad \text{but} \quad S \not> \frac{3}{4}(d_b - d_n)$$

where $K = 2{\cdot}5\, A_{stc}/A_{sta}$, d_b = depth of bottom layer of longitudinal tension reinforcement, $d_b - d_1$ for one layer of bars, d_n = depth of concrete in compression calculated by the load factor method CP 114, A_{stc} = minimum area of longitudinal tension steel required to resist the bending tensile stresses at the inclined section under consideration as calculated by the load factor method, and A_{sta} = actual area of longitudinal tension steel crossing the inclined section under consideration.

With continuous beams it is necessary to consider stopping off points as well as the support sections. The B.R.S. tests have shown that deformed bars may safely be stopped off in accordance with the bending moment diagram if allowance is made for the inclined cracking that may occur. Smooth bars should not be stopped off in a zone of longitudinal tension.

Equation 5.1 will be applied to a section subjected to combined shear and bending and compared with CP 114 recommendations. A further comparison will be made with the A.C.I. building code which gives recommendations for ultimate strength design with regard to shear reinforcement.

Example 5.1

A continuous reinforced concrete beam *(Figure 54)* is subjected to a total load of 3·5 kips/ft. run. Given the following design data proportion the penultimate support section in accordance with CP 114:1957 and on an ultimate load basis assuming a factor of safety of 2·0.

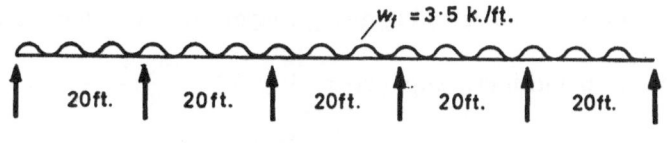

Figure 54

Design Data

C_u	= 4,500 lb./in.^{2}	f_y	= 60,000 lb./in.^{2}
p_{cb}	= 1,500 lb./in.^{2}	w_d	= 2·3 k./ft.
p_{st}	= 30,000 lb./in.^{2}	w_s	= 1·2 k./ft.
b	= 15 in.		

From Table 15, CP 114 the penultimate support moment

$$= \frac{w_d L^2}{10} + \frac{w_s L^2}{9}$$

$$= \frac{2·3 \times 20^2}{10} + \frac{1·2 \times 20^2}{9}$$

$$= 92·0 + 53·3 = 145·3 \text{ k. ft.}$$

Using the load factor method $\quad RM = \frac{P_{cb}}{4} bd^2$

thus
$$145·3 \times 10^3 \times 12 = 375 \times 15 \times d^2$$
$$d = 17·6 \text{ in.}$$

Steel area
$$A_{st} = \frac{145·3 \times 12}{0·75 \times 17·6 \times 30,000}$$
$$= 4·4 \text{ in.}^2$$

Adopt
$$4 - 1\tfrac{1}{8} \text{ in. dia} = 3·98 \text{ in.}^2$$

$$1 - \tfrac{3}{4} \text{ in. dia} = 0·44 \text{ in.}^2$$
$$A_{st} = 4·42 \text{ in.}^2$$

These bars can be placed in a single layer, for a width of 15 in.

Shear at penultimate support $= 3 \cdot 5 \times 10 + \dfrac{145 \cdot 3}{20}$

$$= 35 \cdot 0 + 7 \cdot 27$$
$$= 42 \cdot 27 \text{ kips}$$

Assuming $4\frac{1}{2}$-in. dia. legs $(A_w = 0 \cdot 785)$ and mild steel stressed to 20,000 lb./in.², then the stirrup spacing in accordance with CP 114 will be

$$s = \frac{A_w l_a f_{st}}{Q}$$

where Q = shear force at section considered, l_a = lever arm, s = stirrup spacing and f_{st} = permissible stress in shear reinforcement.

$$s = \frac{0 \cdot 785 \times 0 \cdot 75 \times 17 \cdot 6 \times 20{,}000}{42 \cdot 27 \times 10^3} = 4 \cdot 9 \text{ in.}$$

Adopting the B.R.S. recommendations and assuming $f_y = 40{,}000$ lb./in.² then for

$$Q_u = 2 \times 42 \cdot 27 \quad \text{and} \quad d_n = 0 \cdot 5 \times 17 \cdot 6$$
$$L = K(d_b - d_n)$$
$$= K(17 \cdot 6 - 0 \cdot 5 \times 17 \cdot 6)$$
$$= K\,8 \cdot 8$$

At the support section $A_{stc} = A_{sta}$ thus

$$K = 2 \cdot 5 - 1 = 1 \cdot 5$$
$$L = 1 \cdot 5 \times 8 \cdot 8 = 13 \cdot 2$$

Thus from equation 5.1 $= \dfrac{9 \times 0 \cdot 785 \times 13 \cdot 2 \times 40{,}000}{2 \times 42 \cdot 27 \times 10^3} = 4 \cdot 4 \text{ in.}$

The B.R.S. equation gives a slightly reduced spacing of stir-rups. This value will now be compared with the A.C.I. re-commendations. According to the A.C.I. code the nominal ultimate shear stress is computed from the equation

$$V_u = \frac{V_u}{bd} \quad \text{but} \quad v_u \ngtr \phi \cdot 10(C_c)^{\frac{1}{2}} \tag{5.2}$$

Where v_u = nominal ultimate shear stress, V_u = total ulti-mate shear, b = width of compression face of flexural member, and d = effective depth of section.

Thus $$v_u = \frac{2 \times 42 \cdot 27}{15 \times 17 \cdot 6} = 321 \text{ lb./in.}^2$$

The maximum shear can be considered as that at the section a distance d from the face of the support, but for this example the above value will be assumed.

The shear stress permitted on an unreinforced web shall not exceed that given by

$$v_c = \phi \left(1 \cdot 9(C_c)^{\frac{1}{2}} + 2{,}500 \frac{pVd}{M} \right)$$

where ϕ = capacity reduction factor = 0·85 for shear

C_c = cylinder strength $= 0\cdot7\ C_u = 3{,}150\ \text{lb./in.}^2$

p = steel percentage $= 4\cdot4/17\cdot6 \times 15 = 0\cdot0167$

V = total shear at section $= 42\cdot27$ kips

M = bending moment $= 145\cdot3$ kip.ft.

v_c = shear stress carried by concrete

thus $$v_c = 0\cdot85 \left(1\cdot9 \times 56\cdot1 + \frac{2{,}500 \times 0\cdot0167 \times 42\cdot27 \times 10^3 \times 17\cdot6}{145\cdot3 \times 10^3 \times 12} \right)$$

$$= 0\cdot85(106\cdot4 + 17\cdot8)$$

$$= 105\cdot5 \text{ lb./in.}^2$$

except that v_c shall not exceed $3\cdot5\phi(C_c)^{\frac{1}{2}} = 167$ lb./in.². If the value v_u exceeds v_c then the excess must be carried by shear reinforcement which for vertical stirrups are computed by

$$A_w = V'_u S / \phi f_y d$$

where A_w = area of web reinforcement, f_y = yield strength of reinforcement, and V_u = ultimate shear carried by web reinforcement.

$$V'_u = 15 \times 17 \cdot 6(321 - 105 \cdot 5)$$

$$= 56 \cdot 9 \text{ kips}$$

$$S = \frac{A_w \phi f_y d}{V'_u}$$

$$= \frac{0 \cdot 735 \times 0 \cdot 85 \times 40 \times 10^3 \times 17 \cdot 6}{56 \cdot 9 \times 10^3} = 8 \cdot 27 \text{ in.}$$

The A.C.I. recommendations give a stirrup spacing of almost double the B.R.S. equation for this example. However the B.R.S. tests have shown that a factor of safety of at least $1 \cdot 7$ against a shear failure occuring before a flexural failure is obtained by the use of equation 5.1 with the exception of a beam in which smooth bars were stopped off in accordance with the bending moment diagram. The A.C.I. code gives recommendations for the spacing of inclined bars which was not included in the B.R.S. tests.

For the ultimate load design the Russian recommendations will be assumed for which the penultimate support moment is $w_t L^2 / 11$.

Ultimate moment $= 2 \times 3 \cdot 5 \dfrac{20^2}{11} 12$

$= 3055 \text{ k.in.}$

For $C_u = 4{,}500 \text{ lb./in.}^2$ $M_u = \dfrac{4{,}500}{6} 15 d^2$

Thus
$$d^2 = \frac{3,055 \times 10^3}{4,500 \times 2 \cdot 5}$$

$$d = 16 \cdot 5 \text{ in.}$$

$$A_{st} = \frac{3,055 \times 10^3}{0 \cdot 75 \times 16 \cdot 5 \times 60,000}$$

$$= 4 \cdot 12 \text{ in.}^2$$

Compared with the previous result the ultimate load design for bending gives a reduced depth and steel area. The shear reinforcement will be calculated in accordance with the B.R.S. equation.

$$Q_u = 2 \times 35 + \frac{3055}{20 \times 12}$$

$$= 70 + 12 \cdot 74 = 82 \cdot 74 \, k$$

$$L = K(16 \cdot 5 - 0 \cdot 5 \times 16 \cdot 5)$$

$$= K8 \cdot 25$$

At the support section $A_{sta} = A_{stc}$ thus

$$K = 2 \cdot 5 - 1 = 1 \cdot 5$$

$$L = 1 \cdot 5 \times 8 \cdot 25 = 12 \cdot 38$$

From equation 5.1
assuming $A_w = 0 \cdot 785$
$$s = \frac{0 \cdot 9 \times 0 \cdot 785 \times 12 \cdot 38 \times 40,000}{82 \cdot 74 \times 10^3}$$

$$= 4 \cdot 24 \text{ in.}$$

Suppose that the effective depth of 17·6 in. was maintained for the ultimate load design. The steel area for bending can be obtained by use of design charts prepared by L. L. Jones[17]

$$\frac{f_y}{C_u} = \frac{60,000}{4,500} = 13 \cdot 33$$

$$\frac{M_u}{bd^2 f_y} = \frac{3,055 \times 10^3}{15 \times 17 \cdot 6^2 \times 60,000} = 0 \cdot 011$$

From the design chart for singly reinforced sections

$$p = 0.013\ 7$$

thus

$$A_{st} = 0.013\ 7 \times 15 \times 17.6$$
$$= 3.62 \text{ in.}^2$$

Using equation 5.1 for calculating the shear reinforcement

$$L = K(17.6 - d_n)$$

d_n can be determined by considering the longitudinal equilibrium of the section

$$\frac{4}{9}\ 4{,}500 \times 15\ d_n = 60{,}000 \times 3.62$$

thus

$$d_n = 7.25 \text{ in.}$$
$$L = K(17.6 - 7.25)$$
$$= 1.5 \times 10.35 = 15.5$$

thus for $A_w = 0.785$,

$$s = \frac{0.9 \times 0.785 \times 15.5 \times 40{,}000}{82.74 \times 10^3}$$

$$= 5.29$$

Thus an ultimate load design which is not based on full utilization of the compression zone ($d_n < d/2$) should produce economy in steel both for shear and bending.

RESEARCH ON SHEAR FAILURE

Extensive tests have been carried out in Germany[18] with a view to supporting proposals for the design of shear reinforcement on an ultimate load basis. Recommendations for the design of reinforced concrete beams subjected to combined shear and bending have also been published by C. Erdei[19]. This book indicates the limitations of CP 114 and the writer aims to give a theory for the effects of combined bending and shear and also proposals for practical design. In an introductory book of this nature it is not possible to cover in detail all shear failure theories, but if extensive redistribution of moment is to be assumed then it is recommended that shear reinforcement calculations are based on the B.R.S. equation as it gives an ample factor of safety against shear failure occuring before flexural failure. The bond stresses at ultimate load must also be checked. For the previous example the bond stress for an ultimate shear force of $2 \times 42 \cdot 27$ kips may be checked from the formula given in CP 114.

$$f_B = \frac{Q}{l_a 0}$$

Where 0 = sum of perimeters of bars in the tensile reinforcement

thus
$$f_B = \frac{2 \times 42 \cdot 27 \times 10^3}{0 \cdot 75 \times 17 \cdot 6(4 \cdot 3 \times 54 + 2 \cdot 36)}$$

$$= 388 \text{ lb./in.}^2$$

assuming an ultimate bond stress of $\dfrac{10(C_c)^{\frac{1}{2}}}{D}$ then

$$f_{b,\,\text{ult.}} = \frac{10 \times 56 \cdot 1}{1 \cdot 125} = 498 \text{ lb./in.}^2$$

This is well in excess of the actual value at ultimate load. Tests have shown that when stirrups are included in beams[20] the bond stresses for deformed bars are very much increased. For $\frac{3}{4}$ in. diameter deformed bars in a beam with concrete of crushing strength in the order of 4,500 lb./in.^{2} the average stress at ultimate load was about 1,000 lb./in.^{2}

Torsion

The combination of bending, shear and torsion is not covered in CP 114 and the A.C.I. code. The most practical recommendations are given in the Australian code for concrete

in buildings, which are the result of work carried out by
H. J. Cowan[21].

It should be remembered that shear and torsional stresses
can be additive. The shear stress may be calculated in accordance with the expression $f_{sh} = Q/l_a b$.

According to the Australian code the maximum stresses due
to torsion may be computed from the following formulae.

$$\text{For a rectangular section } f_{to} = \frac{5M_t}{x^2 y} \qquad (5.5)$$

$$\text{For T, L and I sections } f_{to} = \frac{3M_t b}{\sum x^3 y} \qquad (5.6)$$

Where $x =$ smaller overall dimension of rectangle, $y =$ larger
overall dimension of rectangle, $\sum x^3 y =$ the sum of the quan-
tities $x^3 y$ of the component rectangles, and $M_t =$ torsional
moment on section.

For a rectangular section the sum of the shear and torsional
stress f_{to} and f_{sh} will be a maximum at the middle of the longer
side. This value should be limited to those given in the B.R.S.
tables for shear stress. Thus for a beam with a cube strength
of 4,500 lb./in.^{2} with no shear reinforcement the value of f_{sh}
and f_{to} is limited to 55 lb./in.^{2} For a beam with shear reinforcement $f_{sh} + f_{to}$ is limited to 300 lb./in.^{2} The Australian code allows the torsional reinforcement to be calculated for the excess
moment above that producing the nominal shearing stress.
To conform with British practice the reinforcement would be
calculated for the total moment if it produces a shear stress in
excess of the nominal value. The quantity of reinforcement required may be computed from the following expressions according to the Australian code.

$$\text{Area of closed hoops (two legs) } A_{sv} = \frac{SM_t}{8} x' \cdot y' f_s \qquad (5.7)$$

Area of longitudinal steel $\qquad A_s = A_{sv} \dfrac{x' + y'}{S}$ (5.8)

Where x' = smaller dimension of closed loop, y' = larger dimension of closed loop, f_s = stress in shear reinforcement, S = spacing of loops, and M_t = torsional moment, the total value to conform with British practice for calculating shear reinforcement.

At least one longitudinal bar should be placed in each corner of the loops. Care must be taken that adequate anchorage is obtained for the closed loops. This reinforcement is in addition to that required for normal shear. The above procedure does not represent an ultimate load design but provided the B.R.S. recommendations for shear stresses are adhered to, the commencement of large scale cracking should be prevented and a sudden failure will be precluded. If torsion dominates the design of a concrete beam then it may be prudent to avoid an ultimate load design and a more accurate assessment of the torsional stresses will be required[21] than those given by the equations in the Australian code.

PRESTRESSED CONCRETE

Shear

The A.C.I. code recommendations for prestressed concrete base the design of shear reinforcement on ultimate strength and the area of shear reinforcement should not be less than

$$A_v = \frac{(V_u - \phi V_c)S}{\phi d f_y}$$ (5.9)

nor less than $\qquad A_v = \frac{A_s}{80} \cdot \frac{f'_s}{f_y} \frac{S}{d} \left(\frac{d}{b'}\right)^{\frac{1}{2}}$ (5.10)

Where A_v = area of web reinforcement, ϕ = capacity reduction factor = 0·85, V_c = shear carried by concrete, S = spacing of web reinforcement, d = distance from extreme fibre to centroid of prestressing force, f_y = yield stress of web reinforcement, A_s = area of prestressed tendons, f_s' = ultimate strength of prestressing steel, and b' = minimum width of web of flanged member.

The shear V_c at diagonal cracking is taken as the lesser of V_{ci} and V_{cw} which for nominal density concrete are given by

$$V_{ci} = 0 \cdot 6\, b'd(C_c)^{\frac{1}{2}} + \frac{M_{cr}}{\dfrac{M}{V} - \dfrac{d}{2}} + V_d \qquad (5.11)$$

but not less than $1 \cdot 7\, b'd(C_c)^{\frac{1}{2}}$

Where M_{cr} = net flexural cracking moment, M = bending moment due to applied loads, V = shear due to applied loads, and V_d = shear due to dead load.

M and V relate to external ultimate loads acting on the member excepting those applied to the member by the prestressing tendons. V_{ci} = shear at diagonal cracking due to all loads when such cracking is the result of combined shear and bending.

$$M_{cr} = \frac{I}{y} \cdot \left(6(C_c)^{\frac{1}{2}} + f_{pe} - f_d \right) \qquad (5.12)$$

I = second moment of area of section, y = distance from centroid of section to extreme fibre in tension, f_d = stress due to dead load at extreme fibre of a section at which tension stresses are caused by applied loads, f_{pe} = compressive stress in concrete due to prestress only, after losses, at the extreme fibre of a section at which tension stresses are caused by applied loads,

and
$$V_{cw} = b'd \left(3 \cdot 5\,(C_c)^{\frac{1}{2}} + 0 \cdot 3 f_{pc} \right) + V_p \qquad (5.13)$$

where V_{cw} = shear force at diagonal cracking due to all loads when such cracking is the result of excessive principal tensile stresses in the web, V_p = vertical component of the effective prestress force at the section considered, and f_{pc} = compressive stress in the concrete, after losses, at the centroid of the cross section resisting the applied loads, or at the junction of the web and flange when the centroid lies in the flange.

In the British code of practice for prestressed concrete CP 115:1959 design of shear reinforcement is based on limiting the principal tensile stresses at working and ultimate load for uncracked sections.

The values are given below.

Specified works cube strength for concrete (lb./in.²)	P.T.S. at working load (lb./in.²)	P.T.S. at ultimate load (lb./in.²)
4,500	125	300
6,000	150	350
7,500	175	400

The principal tensile stress due to shear, bending and prestress is calculated as below. Using the notation given in *Figure 55* the shear stress at any level $x-x$ is given by

$$f_{sh} = \frac{SA\bar{y}}{Ib} \qquad (5.14)$$

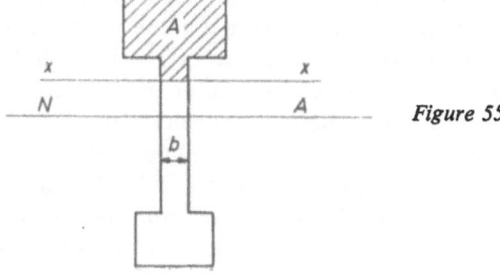

Figure 55

Where S = total shear at section considered, I = second moment of area of whole section about centroid N−A, $\bar{y}$ = distance of centroid of area A from N−A, b = width of section A_t level $x-x$.

The stress at level $x-x$ due to the prestress force P is given by

$$f = \frac{P}{A} \pm \frac{Pey}{I}$$

and the stress due to bending moment M at the section considered is given by

$$f_b = \frac{My}{I}$$

let $$f_b + f_p = f_1$$

then using the principal stress formula

$$F = \frac{f_1}{2} - \left\{ \left(\frac{f_1'}{2} \right)^2 + f_{sh}^2 \right\}^{\frac{1}{2}} \tag{5.15}$$

where F = Principal tensile stress.

If the prestressing tendons are inclined to the longitudinal axis of the beam then the shear force S due to the applied loading will be reduced by an amount equal to the vertical component of the prestressing force at the section considered.

If the limiting principal tensile stress at working load is exceeded then shear reinforcement should be provided in accordance with clause 311 of CP 115. This varies linearly with the principal tensile stress from a value of zero for the stress given in the table to 1·0 for a stress of 1·5 times that given. When the principal tensile stress is greater than 1·5 times the stress given in the table the whole of the shear force should be carried by reinforcement. Where the principal tensile stress at ultimate load exceeds that given in the table the whole of the shear, in excess of that resisted by the tendons inclined to the longitudinal axis, should be resisted by shear reinforcement acting at a stress not exceeding 80 per cent of the yield stress f_y (or 0·2 per cent proof stress, where appropriate). The code also states that special consideration should be given to the shear resistance under ultimate load conditions where the section is cracked in bending. The area of shear reinforcement required at ultimate load, if the limiting principal tensile stresses is exceeded, could be expressed as follows to comply with CP 115 (notation as A.C.I. code)

$$A_v = \frac{V_u S}{0·8 f_y d} \qquad (5.16)$$

This is of similar form to equation 5.9, the difference being that equation 5.9, takes into account the shear resistance of the concrete whereas the use of equation 5.16 means that all the shear is taken by the web reinforcement. In order to prevent the development of large cracks, calculated principal tensile stresses in much excess of the limiting values in CP 115 should be avoided, irrespective of the amount of shear reinforcement employed.

The use of thin webs is more common in prestressed than in reinforced concrete members and thus the use of shear reinforcement is desirable even when the principal tensile stresses are low. Recommendations for reinforcement in thin webs are given in clause 315 of CP 115 with a maximum spacing of the shear web depth. With very thin webs there is a possibility of failure due to high principal compressive stress, which would occur with some violence even though the web is heavily reinforced.

Torsion

As with reinforced concrete the combination of bending shear and torsion is not covered in CP 115 or the A.C.I. code. The torsion provisions in the Australian Building Code give an upper limit of 3 $(C_c)^{\frac{1}{2}}$ for principal stresses at working load irrespective of the amount of untensioned reinforcement required. The code also gives limitations to the principal tensile stress at ultimate conditions. If the torsion is a dominant factor in design then calculated ultimate principal tensile stress should not exceed $4(C_c)^{\frac{1}{2}}$ irrespective of the amount of untensioned shear reinforcement. The term 'dominant factor' is not clearly defined, but it is indicated that a relatively small bending moment on a section should prevent a catastrophic failure. If the torsion is not a dominant design factor, then if the calculated ultimate principal tensile stress due to torsion exceeds the calculated ultimate principal tensile stress due to transverse shear, torsional shear reinforcement is provided for that part of the ultimate twisting moment which produces a calculated principal tensile stress greater than $4(C_c)^{\frac{1}{2}}$. In addition reinforcement is provided for the whole of the transverse shear (the shear resistance of the concrete is ignored for calculation of transverse shear reinforcement that is V_c is put equal to zero in equation 5.9).

When the calculated ultimate principal tensile stress due to torsion does not exceed the calculated ultimate principal tensile stress due to transverse shear, reinforcement is provided for the whole of the ultimate torsional moment, but due allowance is made for the ultimate resistance of the concrete to shear, that is V_c is not put equal zero in equation 5.9.

The area of torsional shear reinforcement may be calculated from equations 5.7 and 5.8. For ultimate conditions the steel stress f_s is replaced by the yield stress f_y. For further information of the application of the torsion provisions of the Australian prestressed concrete code the reader is referred to H. J. Cowan's book[21] which gives useful practical information on the subject of torsion in concrete.

6

FACTOR OF SAFETY AND SERVICEABILITY

SAFETY

The design of reinforced concrete structures on an elastic basis bears little relation to its actual behaviour under load and gives no true indication of the margin of safety against failure. This may be expressed in terms of a load factor which may be defined as the ratio of the ultimate load to the working load.

This is a simple definition but has far-reaching implications. The first problem is to assess the working load and the second is to determine the factor by which the working load should be multiplied to give the ultimate load. In CP 114:1957 it is stated that the resistance moments of beams and slabs may be calculated to have a load factor generally of 2·0. In view of the greater variability of the strength of concrete compared with that of steel a fictitious value of two-thirds the actual cube strength is adopted for the cube strength of concrete in the calculations. In the amendment to CP 114 issued in February 1965 the load factor is reduced to 1·8. However, a distinction is made between nominal and designed concrete mixes. For nominal mixes ultimate strength calculations are based on a cube strength of three-fifths the actual cube strength and for designed concrete mixes the cube strength is taken as two-thirds the actual cube strength.

Further, for deformed high tensile steel reinforcing bars of diameter not greater than $\frac{7}{8}$ in. the permissible stress of 0·55 the guaranteed yield or proof stress implies a reduction of load factor to about 1·8. It is unfortunate that CP 114 gives no recommendations regarding the ultimate load approach to shear and bond. In fact it is possible to design a concrete member in accordance with CP 114 that will have a factor of safety against shear failure considerably less than that in bending as indicated in the previous chapter.

The A.C.I. building code (318 − 63) stipulates that for structures in which the effects of wind and earthquake can be neglected the ultimate load should be at least equal to

$$1·5 \text{ dead load} + 1·8 \text{ live load.}$$

The use of different load factors for dead and live load is more rational than the CP 114 approach of 1·8 times the total load as the dead load can be assessed fairly accurately, whereas the live load can be subject to considerable variation.

With regard to the assessment of working loads the designer normally relies on the relevant code of practice for loading which in England is CP 3—Chapter V (1952), Loading.

It is a matter of conjecture whether the tables of superimposed loads for various types of floors do in fact represent the true imposed or live loading.

Consider for example the minimum imposed load of 50 lb./ft.^{2} for office floors above the entrance floor. During the last ten years considerable changes have taken place with regard to the planning of office floors such as the number of persons occupying a given area, office equipment and furniture etc., which mean that modification of the above loading may be necessary. Research into the assessment of working loads lags behind our knowledge of the behaviour of structural materials under load. There is an obvious need for reassessment of working loads if accurate load factors are to be realized.

The numerical value of the load factor should be related to the function and useful life of the structure. The load factor for a motorway bridge should obviously be greater than that for a temporary shelter for equipment on a building site. The Ministry of Transport requirement for load factor is 1·5 dead load plus 2·5 live load and it is implied that the useful life of a bridge should be well over 100 years. The useful life of an office block would probably be about half that of a bridge. It should be emphasized that due to deterioration of the structure caused by factors such as spalling of the concrete, corrosion of reinforcement etc., the load factor will reduce with increase in age of the structure, thus care in detailing and maintenance of cover to reinforcement etc. are essential if durability is to be ensured.

A report on structural safety[22] issued by the Institution of Structural Engineers relates the load factor to considerations such as human risk, economic effects of failure, useful life of structure workmanship etc. and is a step towards a rational view of the problem of structural safety.

Too much reliance should not be made on calculations with regard to structural safety. A bad detail can invalidate the most sophisticated calculation. Many structures are designed without adequate consideration of the effects of creep, shrinkage and thermal movement. These effects may produce cracking which will seriously affect the durability of a structure. An even distribution of steel throughout a structure and the avoidance of stress concentrations due to bad proportioning of reinforcement are essential if the useful life of the structure is to be ensured. A typical example of serious cracking which can occur due to bad detailing and lack of consideration to thermal and shrinkage effects is the trouble which is encountered with reinforced concrete parapet beams. Although relatively lightly loaded, these beams are often badly cracked. Additional reinforcement is required to counteract shrinkage and temperature stresses which may be in excess of those induced by dead and superimposed loading.

Finally the importance of good detailing cannot be overemphasized. Splitting of the concrete cover to longitudinal bars in the tensile zone of a beam may be counteracted by transverse reinforcement located between the bar and the concrete surface. The layout of the transverse reinforcement is important[23] if it is to resist splitting in vertical and horizontal planes. *Figure 56* indicates how the layout of transverse reinforcement influences the splitting of the concrete cover. Further, tests on beams with no stirrups[20] indicated that the bond stress for de-

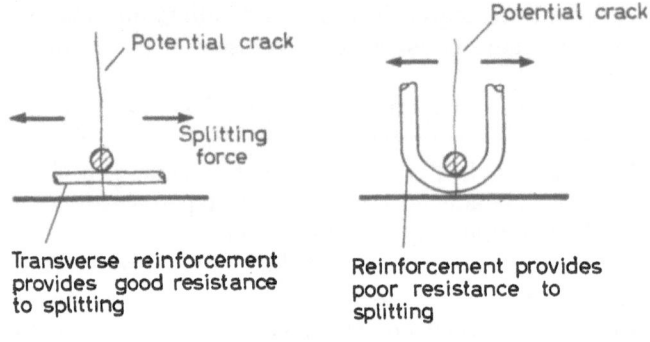

Figure 56

formed bars was much higher than for smooth bars but the concrete required reinforcement to prevent bursting due to radial forces caused by the bar ribs after movement of the bar had taken place. The provision of nominal stirrups appears to be adequate to prevent bursting and also has the effect of nearly doubling the value of the average bond stress at ultimate load.

SERVICEABILITY

The serviceability of a reinforced concrete structure is related to its performance at working load. This is generally assessed in terms of deformation and crack width. Excessive deformation of beams and slabs may damage finishes and partitions and the psychological effects of a sagging member should also be considered even though the strength of the member is not impaired.

Calculations for deformations of reinforced concrete structures are not reliable as it is difficult to assess the true stiffness of a beam and slab system, the effect of cracking on the stiffness, and the creep deformations. Calculations based on the standard equation

$$\Delta = K \frac{WL^3}{EI}$$

(where Δ = deformation, K = constant depending on end conditions of member, W = total load on span, L = span of member, E = modulus of elasticity, and I = second moment of area) give little indication of actual deformations and thus it is normal practice in England to keep within certain span-to-depth ratios. CP 114 gives permissible span-to-depth ratios for beams as follows.

Beam Type	$\dfrac{Span}{Depth}$
Simply supported	20
Continuous	25
Cantilever	10

It is further stipulated that for members with steel stresses greater than 20,000 lb./in.^{2} or concrete stresses greater than 1,500 lb./in.^{2} the span-to-depth ratio should not exceed 90 per cent of the above values. For members with steel stresses greater than 20,000 lb./in.^{2} and concrete stresses greater than 1,500 lb./in.^{2}, the span-to-depth ratio should not exceed 85 per cent of the above values. If deflection calculations are made, then creep strains should be considered, which reduce the effective modulus of elasticity of the concrete to half or one third of its initial value.

The A.C.I. code gives span-to-depth ratio as below, if deflections are not computed.

Beam	$\dfrac{Span}{Depth}$
Simply supported	20
One end continuous	23
Both ends continuous	26
Cantilever	10

Recommendations for calculating immediate and long term deflections are also given.

The calculation of crack widths is not covered in CP 114. However, crack width formulae have been proposed by several research workers,[24] and the C.E.B. draft recommendations for an international code of practice give the following limits which may generally be accepted for maximum widths of cracks.

For internal structural parts in a normal atmosphere 0·3 mm
For internal structural parts in a humid or aggressive
  atmosphere and external structural parts exposed
  to the weather   0·2 mm
For internal or external structural parts exposed to a
  particularly aggressive medium or where water
  tightness is needed   0·1 mm

The A.C.I. code gives full scale test values of 0·015 in. (0·38 mm) for internal members and 0·01 in. (0·25 mm) for external members. These are average crack widths. Cracking may have a considerable influence on the durability of a structure and the possibility of corrosion must be investigated. Tests car-

ried out by P. W. Abeles[25] on beams bent to such an extent
that a hair crack of 0·1 mm was induced and exposed to an
extremely corrosive environment for a period of one year, indicated that for covers of 1 in. and $\frac{1}{2}$ in. to 0·2 in. diameter
wires, no significant corrosive effect occured. Further it is the
density of the concrete rather than the depth of cover that is
important with regard to corrosion. The crack width formulae
relate to cracking normal to the direction of tension (longitudinal) reinforcement and do not take into account cracks paral-
lel to the reinforcement. These cracks will be more significant
with regard to corrosion and can be counteracted by the provision of transverse reinforcement and ensuring adequate compaction of the concrete. Formulae for calculating crack widths
can be expressed in the form

$$\Delta = \frac{K}{10} \frac{\phi f}{p} \tag{6.1}$$

Where Δ = crack width (mm), K = coefficient, f = tensile
stress in steel in crack (kg/mm²), and ϕ = bar diameter (mm).

The above formula is a simplified version of expressions developed by research workers[24] and the coefficient K may be
expressed in the following form

$$K = \frac{100k_1 t}{E_s f_b} \tag{6.2}$$

where k_1 = dimensionless constant depending on the relation
for bond between the concrete and steel, t = tensile strength
of the concrete (kg/cm²), f_b = maximum bond stress (kg/cm²),
E_s = modulus of elasticity of steel (kg/mm²).

Values of k_1 vary from 0·5 to 1·0. Values of K for smooth and
deformed bars have been tabulated[24] and there is considerable
variation depending on which of the theories is adopted. The
use of equation 6.1 will be illustrated by two examples, one in
which deformed bars are stressed to the permissible value of
33,000 lb./in.² (CP 114) and the other in which the bars are

stressed to a value well in excess of that permitted in CP 114. The results obtained will be compared with measured values obtained from tests on beams with deformed bar reinforcement[20] which give the following results.

Steel stress as a proportion of working stress of 33,000 lb./in.²	Average maximum crack width in inches
1·0	0·004 1 (0·1 mm)
1·25	0·005 0 (0·13 mm)
1·50	0·007 2 (0·18 mm)
1·75	0·008 5 (0·22 mm)

In the tests the bottom cover of the main steel was 1 in. and the bar diameter $\frac{3}{4}$ in.

Example 6.1

An internal reinforced concrete beam with 4-No $\frac{3}{4}$ in.dia. high tensile deformed bars *(Figure 57)* has a width of 10 inches and cover of 1 inch. If the bars are stressed to 33,000 lb./in.² determine the crack width.

Steel area (4-$\frac{3}{4}$ in. dia)   = 1·767 in.²

Steel stress   = 33,000 lb./in.²

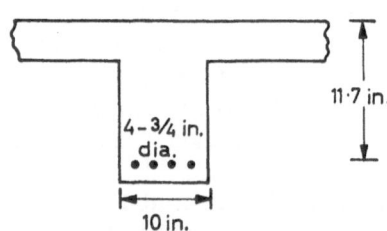

Figure 57

Steel percentage
$$= \frac{1 \cdot 767 \times 100}{10 \times 11 \cdot 7}$$
$$= 1 \cdot 51$$

From equation 6.1

$$\Delta = \frac{K}{10} \cdot \frac{\phi f}{p}$$

$$\phi = 0 \cdot 75 \times 25 \cdot 4 \text{ mm}$$

$$f = 33{,}000 \text{ lb./in.}^2 = 0 \cdot 000 \, 703 \text{ kg/mm}^2$$

The tabulated values of K vary considerably and the maximum value quoted is $2 \cdot 5 \times 10^{-3}$: for this value:

$$\Delta = 29 \cdot 3 \times 2 \cdot 5 \times 10^{-3}$$
$$= 0 \cdot 073 \text{ mm.}$$

A modified version of equation 6.1 has been proposed, which includes a parameter for the influence of the steel area to the area of concrete immediately surrounding it.*†

* The following crack width formula is given in an Institution of Structural Engineers publication ('Formula for the computation of the strength of reinforced and prestressed concrete members', March, 1966).

$$\text{Maximum crack width} = \left(4 \cdot 5 + \frac{0 \cdot 4}{r_e} \right) D_r \frac{f_{st}}{c}$$

where D_r = diameter of reinforcement, f_{st} = stress in steel in tension, and $c = 47 \cdot 5 \times 10^6$ lb./in.^{2} for deformed bars $29 \cdot 5 \times 10^6$ lb./in.^{2} for plain bars

$$r_e = \frac{A_{st}}{A_{et}} \quad \text{where} \quad 0 \cdot 02 \leqslant r_e \leqslant 0 \cdot 2$$

A_{st} = area of tensile reinforcement and A_{et} = area of tensile concrete zone having same centroid as that of the steel.

† A crack width formula has recently been developed by the Cement and Concrete Association which will shortly be made available in a Technical Report. This formula indicates that crack widths are primarily influenced by the concrete cover and the stress in the reinforcement.

The calculated value of the crack width, using the maximum value of K is only 73 per cent of the measured value for the same steel stress of 33,000 lb./in.[2]

Example 6.2

The beam considered in example 6.1 has been designed on a plastic basis for a penultimate support moment of $w_t L^2/12$. If the load factor is 1·8 and the ratio of live to dead load 0·5, determine the maximum crack width at working load in accordance with equation 6.1. Yield stress for deformed bars is 60,000 lb./in.[2]

From Figure 7 (Chapter 1) for $w_s/w_d = 0·5$ the penultimate support moment is $w_t L^2/9·6$ for an elastic design. Thus for a factor of safety of 1·8 the steel stress at working load, assuming an approximately elastic distribution of moments at working load, is

$$f_s = \frac{60,000}{1·8} \times \frac{12·0}{9·6}$$

$$= 41,700 \text{ lb./in.}^2$$

crack width $\qquad W = \frac{K}{10} \frac{\phi f}{p}$

$$= \frac{2·5}{10^4} \times 0·75 \times 25·4 \times 41,700 \times$$

$$\times 0·000\ 703$$

$$= 0·14 \text{ mm.}$$

The steel stress of 41,700 lb./in.[2] is equivalent to $\dfrac{41,700}{33,000} =$ 1·2×6 of the working load stress which corresponds to a measured crack width[20] of 0·13 mm. This is an acceptable crack width except possibly for an extremely corrosive environment or where water tightness is required.

To summarize, the proportioning of reinforced concrete beams based on Russian recommendations for moment coefficients would not normally require a check on permissible hinge rotations or on serviceability provided that an under-reinforced section is adopted.

Further redistribution will require serviceability and rotation checks but the calculations are subject to so many variables that their suitability for the design office is in doubt. Shear and bond at ultimate conditions must always be considered.

In view of the extensive research carried out in recent years it is anticipated that the next revision of the British Standards Code of Practice CP 114 will give recommendations for the ultimate load design of continuous concrete beams providing the moment/rotation characteristics are of the bi-linear form.

REFERENCES

1. 'Ultimate load design of concrete structures'. *Proceedings of the Institution of Civil Engineers*. February, 1962 (reprinted together with the discussion as a separate booklet in 1964).

2. *Recommendations for an International Code of Practice for Reinforced Concrete*. Published jointly by the American Concrete Institute and the Cement and Concrete Association.

3. 'Inelastic hyperstatical frames—analysis and application of the international correlated results'. Professor A. L. L. Baker and A. M. N. Amarakone. Presented for discussion at a joint meeting of the Cement and Concrete Association, the Institution of Civil Engineers, the Institution of Structural Engineers and the Reinforced Concrete Association in March, 1965.

4. *Design of Rectangular Beams and Slabs*. Jacques S. Cohen. Published by Concrete Publications Limited.

5. 'The effectiveness of helical binding in the compression zone of concrete beams'. G. D. Base and J. B. Read. *Cement and Concrete Association Technical Report TRA/379*.

6. *The Design of Concrete Structures*. 7th edition. G. Winter and Others. Published by the McGraw-Hill Book Company.

7. *Ultimate Load Theory Applied to the Design of Reinforced and Prestressed Concrete Frames*. A. L. L. Baker. Published by Concrete Publications Limited.

8. *Prestressed Concrete Designers Handbook*. Abeles and Turner. Published by Concrete Publications Limited.

9. *Prestressed Concrete Design and Construction*. 2nd edition. F. Leonhardt. Published by Wilhelm Ernst and Sohn, Berlin, Munich.

10. 'An investigation of the stress distribution in the anchorage zones of post-tensioned concrete members'. J. Zielinski and R. E. Rowe. *Cement and Concrete Association Research Report No. 9.*

11. *The Design and Construction of Prestressed Concrete Structures.* T. Y. Lin. Published by J. Wiley and Sons.

12. 'The collapse method of design being the application of the plastic theory to the design of mild steel beams and rigid frames.' *British Constructional Steelwork Association Publication No. 5,* 1952.

13. *Computer Programme for the Design of Continuous Reinforced Concrete Beams.* A. Weller. Sir Frederick Snow and Partners internal publication.

14. 'Redistribution of design bending moments in reinforced concrete continuous beams'. A. H. Mattock. *Institution of Civil Engineers Paper No. 6314.*

15. 'The strength of statically indeterminate prestressed concrete structures'. Y. Guyon. *Symposium on the Strength of Concrete Structures.* Session C. paper No. 2. Printed by the Cement and Concrete Association.

16. 'Study of shear in reinforced concrete, a new method of proportioning stirrups in beams'. *Department of Scientific and Industrial Research, Building Research Station Note No. B261.*

17. *Ultimate Load Analysis of Reinforced and Prestressed Concrete Structures.* L. L. Jones. Published by Chatto and Windus, London, 1962.

18. 'The Stuttgart Shear Tests 1961'. F. Leonhardt and R. Walther. *Cement and Concrete Association library translation No. 111.*

19. *Design of Reinforced Concrete Beams Subjected to Shear and Combined Bending by the Ultimate Load Theory.* Charles Erdei. London, 1961.

20. 'Tests to determine the behaviour of Tentor steel as tensile reinforcement for concrete'. H. E. Lewis. *Cement and Concrete Association Technical Report TRA/174.* November, 1954.

21. *Reinforced and Prestressed Concrete in Torsion.* H. J. Cowan. Published by Edward Arnold, London. Includes survey of building code requirements.

22. 'Report on structural safety'. Published by the Institution of Structural Engineers. *Structural Engineer*, May 1955.

23. 'Influence of transverse reinforcement on shear and bond strength'. J. R. Robinson. *Journal of the American Concrete Institute*, March 1965.

24. *European Committee for Concrete, information bulletin No. 12.* Issued by the Cement and Concrete Association, London.

25. 'The corrosion of steel in finely cracked reinforced and pre-stressed concrete'. P. W. Abeles and S. Filipek. *Concrete and Constructional Engineering* Vol. LVIII, No. 11, November, 1963.

26. 'The indirect tensile strength of concrete of high compressive strength'. J. D. Dewar. *Cement and Concrete Association, Technical Report No. TRA/377.*

INDEX